Communications in Computer and Information Science 1127

Commenced Publication in 2007
Founding and Former Series Editors:
Phoebe Chen, Alfredo Cuzzocrea, Xiaoyong Du, Orhun Kara, Ting Liu,
Krishna M. Sivalingam, Dominik Ślęzak, Takashi Washio, Xiaokang Yang,
and Junsong Yuan

Editorial Board Members

More information about this series at http://www.springer.com/series/7899

Thuc D. Le · Kok-Leong Ong · Yanchang Zhao ·
Warren H. Jin · Sebastien Wong · Lin Liu ·
Graham Williams (Eds.)

Data Mining

17th Australasian Conference, AusDM 2019
Adelaide, SA, Australia, December 2–5, 2019
Proceedings

 Springer

Editors
Thuc D. Le
School of Information Technology
and Mathematical Sciences
University of South Australia
Adelaide, Australia

Yanchang Zhao
CSIRO Scientific Computing
Canberra, Australia

Sebastien Wong
Consilium Technology
Adelaide, Australia

Graham Williams
Microsoft Proprietary Limited
Singapore, Singapore

Kok-Leong Ong
La Trobe University
Melbourne, Australia

Warren H. Jin
CSIRO Scientific Computing
Canberra, Australia

Lin Liu
School of Information Technology
and Mathematical Sciences
University of South Australia
Adelaide, Australia

ISSN 1865-0929   ISSN 1865-0937   (electronic)
Communications in Computer and Information Science
ISBN 978-981-15-1698-6   ISBN 978-981-15-1699-3   (eBook)
https://doi.org/10.1007/978-981-15-1699-3

Preface

It is our great pleasure to present the proceedings of the 17th Australasian Data Mining Conference (AusDM 2019) held at the University of South Australia, Adelaide, during December 2–5, 2019.

The AusDM conference series first started in 2002 as a workshop initiated by Prof. Simeon Simoff, Dr. Graham Williams, and Prof. Markus Hegland. Over the years, AusDM has established itself as the premier Australasian meeting for both practitioners and researchers in data mining. AusDM is devoted to the art and science of intelligent analysis of (usually big) data sets for meaningful (and previously unknown) insights. Since AusDM 2002, the conference series has showcased research in data mining, providing a forum for presenting and discussing the latest research and developments.

Built on this tradition, AusDM 2019 has successfully facilitated the cross-disciplinary exchange of ideas, experiences, and potential research directions, and pushed forward the frontiers of data mining in academia, government, and industry.

AusDM 2019 received altogether 54 valid submissions, with authors from 16 different countries. The top 5 countries, in terms of the number of authors who submitted papers to AusDM 2019, were Australia (71 authors), China (18 authors), India (14 authors), New Zealand (11 authors), and Germany (5 authors). All submissions went through the double-blind review process, and each paper received at least three peer review reports. Additional reviewers were considered for a clear review outcome, if review comments from the initial three reviewers were inconclusive.

Out of these 54 submissions, a total of 20 papers were finally accepted for publication. The overall acceptance rate for AusDM 2019 was 37%. Out of the 34 Research Track submissions, 11 papers (i.e. 32%) were accepted for publication. Out of the 17 submissions in the Application Track, 8 papers (i.e. 47%) were accepted for publication. Out of the 3 submissions in the Industry Showcase Track, 1 paper (i.e. 33%) was accepted for publication. All the papers from the Industry Showcase Track were invited for oral presentation.

The AusDM 2019 Organizing Committee would like to give their special thanks to Prof. Albert Bifet, Prof. Anton van den Hengel, Dr. Dale Lambert, Prof. Kate Smith-Miles, and Prof. Xin Yao for kindly accepting the invitation to give keynote speeches, and to Dr. Mingming Gong, Dr. Sarah Erfani, Dr. Xingjun Ma, and Ehsan Abbasnejad for organizing and presenting the tutorials. The committee would also like to give their sincere thanks to the University of South Australia for providing admin support and the conference venue ensuring the organization of a successful event. The committee would also like to thank Springer CCIS and the Editorial Board for their acceptance to publish AusDM 2019 papers. This will give excellent exposure of the papers accepted for publication. We would also like to give our heartfelt thanks to all student and staff volunteers at the University of South Australia who did a tremendous job. Last but not least, we would like to give our sincere thanks to all delegates for attending the conference this year and we hope you enjoyed AusDM 2019.

December 2019

Thuc D. Le
Yanchang Zhao
Sebastien Wong
Warren H. Jin
Kok-Leong Ong
Lin Liu
Graham Williams

Organization

Organization Committee

Conference Chairs

Lin Liu   University of South Australia, Australia
Graham Williams   Microsoft, Singapore

Program Chairs

Thuc Le   University of South Australia, Australia
Yanchang Zhao   Data61, CSIRO, Australia
Warren Jin   Data61, CSIRO, Australia
Sebastien Wong   Consilium Technology, Australia

Proceedings Chair

Kok-Leong Ong   La Trobe University, Australia

Publicity Chair

Yee Ling Boo   RMIT University, Australia

Organizing Chairs

Wolfgang Mayer   University of South Australia, Australia
Cristina Garcia   University of South Australia, Australia

Sponsorship Chair

Michael Bewong   University of South Australia, Australia

Web Master

Vu Viet Hoang Pham   University of South Australia, Australia

Steering Committee Chairs

Simeon Simoff   University of Western Sydney, Australia
Graham Williams   Microsoft, Australia

Steering Committee Members

Peter Christen   The Australian National University, Australia
Ling Chen   University of Technology Sydney, Australia
Zahid Islam   Charles Sturt University, Australia

Paul Kennedy	University of Technology Sydney, Australia
Yun Sing Koh	The University of Auckland, New Zealand
Jiuyong (John) Li	University of South Australia, Australia
Richi Nayak	Queensland University of Technology, Australia
Kok–Leong Ong	La Trobe University, Australia
Dharmendra Sharma	University of Canberra, Australia
Glenn Stone	Western Sydney University, Australia
Yanchang Zhao	Data61, CSIRO, Australia

Program Committee

Research Track

Xuan-Hong Dang	IBM T.J. Watson, USA
Philippe Fournier-Viger	Harbin Institute of Technology, China
Ashad Kabir	Charles Sturt University, Australia
Yun Sing Koh	The University of Auckland, New Zealand
Gang Li	Deakin University, Australia
Cheng Li	Deakin University, Australia
Kewen Liao	Charles Darwin University, Australia
Brad Malin	Vanderbilt University, USA
Wolfgang Mayer	University of South Australia, Australia
Veelasha Moonsamy	Deakin University, Australia
Muhammad Marwan Muhammad Fuad	Coventry University, UK
Quang Vinh Nguyen	Western Sydney University, Australia
Hien Nguyen	La Trobe University, Australia
Xuan-Hoai Nguyen	AI Academy, Vietnam
Dang Nguyen	Deakin University, Australia
Ninh Pham	The University of Auckland, New Zealand
Jianzhong Qi	The University of Melbourne, Australia
Yongrui Qin	University of Huddersfield, UK
Mohammad Saiedur Rahaman	RMIT University, Australia
Md Anisur Rahman	Charles Sturt University, Australia
Md Geaur Rahman	Charles Sturt University, Australia
Azizur Rahman	Charles Sturt University, Australia
Jia Rong	Victoria University, Australia
Dharmendra Sharma	University of Canberra, Australia
Xiaohui Tao	University of Southern Queensland, Australia
Dhananjay Thiruvady	Monash University, Australia
Truyen Tran	Deakin University, Australia
Sitalakshmi Venkatraman	Melbourne Polytechnic, Australia
Bay Vo	HUTECH, Vietnam
Kui Yu	Hefei University of Technology, China
Rui Zhang	The University of Melbourne, Australia

Application Track

Hadi Akbarzadeh Khorshidi	The University of Melbourne, Australia
Nathan Brewer	Department of Human Services, Australia
Yonghua Cen	Nanjing University of Science and Technology, China
Adriel Cheng	Defence Science and Technology Group, Australia
Lianhua Chi	La Trobe University, Australia
Yingsong Hu	Department of Social Services, Australia
Ashad Kabir	Charles Sturt University, Australia
Susie Kluth	Department of Social Services, Australia
Dipangkar Kundu	Department of Agriculture and Water Resources, Australia
Sherry Li	Australian Trade and Investment Commission, Australian
Jin Li	Geoscience Australia, Australia
Bin Liang	University of Technology Sydney, Australia
Chao Luo	Department of Health, Australia
Simona Adriana Mihaita	University of Technology Sydney, Australia
Khoa Nguyen	CSIRO, Australia
Shirui Pan	University of Technology Sydney, Australia
Clifton Phua	DataRobot, Singapore
Munir Shah	AgResearch, New Zealand
Yanfeng Shu	CSIRO, Australia
Meina Song	Beijing University of Posts and Telecommunications, China
Ronnie Taib	CSIRO, Australia
Dinusha Vatsalan	CSIRO, Australia
Stephen Wan	CSIRO, Australia
Hongzhi Yin	The University of Queensland, Australia
Huaifeng Zhang	Department of Human Services, Australia

Industry Showcase Track

Rohan Baxter	Australian Taxation Office, Australia
Richard Gao	Department of Health, Australia
Warwick Graco	Australian Taxation Office, Australia
Marcus Suresh	Department of Industry, Australia

Additional Reviewers

Bayu Distiawan
Steven Edwards
Atabak Elmi
Warwick Graco
Haripriya Harikumar
Jiayuan He
Peter Hough
Xinting Huang
Yinhao Jiang
Xiaomei Li
Zhaolong Ling
Guanli Liu
Zhigang Lu

Federico Montori
Do-Van Nguyen
Andrew Perrykkad
Su, Yixin Su
Kaibing Wang
Qinyong Wang
Xiaojie Wang
Yuandong Wang
Khin Nandar Win
Theodor Wyeld
Shuai Yang
Yihong Zhang

Contents

Research Track

Improving Clustering via a Fine-Grained Parallel Genetic Algorithm with Information Sharing

Storm Bartlett[(✉)] and Md Zahidul Islam

School of Computing and Mathematics, Charles Sturt University,
Bathurst 2795, Australia
{sbartlett,zislam}@csu.edu.au

Abstract. Clustering is a very common unsupervised machine learning task, used to organise datasets into groups that can provide useful insight. Genetic algorithms (GAs) are often applied to the task of clustering as they are effective at finding viable solutions to optimization problems. Parallel genetic algorithms (PGAs) are an existing approach that maximizes the effectiveness of GAs by making them run in parallel with multiple independent subpopulations. Each subpopulation can also communicate by exchanging information throughout the genetic process, enhancing their overall effectiveness. PGAs offer greater performance by mitigating some of the weaknesses of GAs. Firstly, having multiple subpopulations enable the algorithm to more widely explore the solution space. This can reduce the probability of converging to poor-quality local optima, while increasing the chance of finding high-quality local optima. Secondly, PGAs offer improved execution time, as each subpopulation is processed in parallel on separate threads. Our technique advances an existing GA-based method called GenClust++, by employing a PGA along with a novel information sharing technique. We also compare our technique with 2 alternative information sharing functions, as well with no information sharing. On 5 commonly researched datasets, our approach consistently yields improved cluster quality and a markedly reduced runtime compared to GenClust++.

Keywords: Genetic algorithm · Clustering · K-Means · Parallel genetic algorithm · Data mining · Machine learning

1 Introduction

Technological progress in recent decades has resulted in the rapid generation and availability of large quantities of data and meta-data in many domains [1]. The challenge of manually deducing clear conclusions from vast stores of data is becoming increasingly more challenging and relevant in today's information-rich world. This explains the accelerating demand for effective and efficent data mining techniques. Clustering has many current and relevant application areas,

© Springer Nature Singapore Pte Ltd. 2019
T. D. Le et al. (Eds.): AusDM 2019, CCIS 1127, pp. 3–15, 2019.
https://doi.org/10.1007/978-981-15-1699-3_1

such as machine learning, image subdivision, corporate enterprise, social networks, medical photography and object identification [2].

K-Means is one of the most popular (unsupervised) clustering algorithms, with several key disadvantages including:

1. Requires the $k - value$ (number of clusters) as input. This is impractical for many users to know in advance and often results in them making an inaccurate guess.
2. It is sensitive to initial seeds that are selected randomly.
3. Its simple hill-climber method for its objective function mean it converges upon local optima, and hence often yields low-quality clusters.

Thus, there exists a demand for clustering approaches that are both simple and avoidant of the pitfalls of K-Means [3], as well as those that actively take advantage of the increasing availability of multiprocessor and/or multicore systems.

1.1   Parallel Genetic Algorithms (PGAs)

Derived from Darwin's Theory of Evolution, genetic algorithms (GAs) are inherently parallel, specifically with regard to their data processing requirements [4].

Our research efforts focus on having separately evolving populations (*subpopulations*), that are each processed on their own thread. Each subpopulation contains a collection of *chromosomes*, where each chromosome represents a full clustering solution (see Figs. 1 and 2) for a particular $k - value$, where k is the number of clusters. That is, each chromosome is a collection of cluster centroids, where each centroid is called a *gene*. Each gene has the same attributes as a data point (or record) of the dataset. Note that in the graphs below, all points in each enclosed region around each centroid belong to whatever centroid is in that region. This collection of clusters make up the complete chromosome.

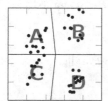   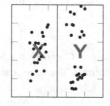   

Fig. 1. Chromosome 1   **Fig. 2.** Chromosome 2   **Fig. 3.** Chromosome 3

PGAs can share information between subpopulations as they evolve to enhance their overall ability to find high-quality solutions to optimization problems. For our particular PGA-based approach, ParallelClust, information sharing occurs periodically (i.e. every X generations) replacing each subpopulation's worst chromosome with another randomly chosen subpopulation's best chromosome. The fitness of any given chromosome is determined using a fitness function. Our research uses the Davies-Bouldin index (DBI) [5], a common evaluation metric for clustering algorithms.

The way the genetic process operates is that it aims to allow each subpopulation to explore its own local optimum, as each subpopulation has its best chromosome stored. Then, by periodically information sharing, it is hoped that good solutions are propagated and combined with other top performers, to find even better solutions. Removing the worst performer is a form of elitism, where we propagate the best solution for the next generation and we remove less fit solutions.

For example, a top performer from one of our subpopulations could be chromosome 1 (Fig. 1), which is then put into the subpopulation that chromosome 2 (Fig. 2) belongs. Then crossover might happen between these 2 chromosomes to create a new top-performing solution. This occurs by taking the first half of the genes of chromosome 2 (X), and combining them with the second half of chromosome 1's genes (B and D), to yield the more optimal solution of chromosome 3 (Fig. 3). It is in this way that GAs generally work, wherein they iteratively improve the solution to any optimization problem over successive generations.

1.2   Main Contributions

We present clear explanation of the components of our approach, justifying them logically. We then empirically compare ParallelClust to its single-population predecessor, GenClust++ [3].

Our 3 main contributions can be summarized as follows:

1. Our new PGA-based clustering approach, ParallelClust, featuring a novel information sharing function and some augmentations that improve performance.
2. An empirical justification of our technique, ParallelClust, using one of 3 information sharing functions. ParallelClust consistently outperforms GenClust++ [3], which uses one large serially-processed population. We demonstrate ParallelClust's improved performance on 5 randomly chosen datasets, containing both numerical and categorical attributes.
3. A graphical presentation of ParallelClust's performance versus GenClust++ [3], showing how the fitness of each population/subpopulation's best chromosome changes over each successive generation.

2   Our Technique: ParallelClust

We use several key components from GenClust++ [3] and combine them with the idea of information sharing between subpopulations. The idea of using a

PGA with periodic information sharing was inspired by research done by HeMI [2]. So overall, our technique could be considered as a parallelised version of GenClust++ [3], featuring various augmentations along with a novel information sharing function. We produce a clustering technique that is superior both in terms of execution time and cluster quality, when compared to its sequentially-processed predecessor, GenClust++ [3].

Each subpopulation has operations such as generation of an initial high-quality population, crossover, information sharing, mutation, chromosome selection and elitism. Our approach consists of 9 components, each individually explained in the following pages.

High-Level Explanation

Algorithm 1 coordinates the overall approach by maintaining a collection of the each subpopulation. Each subpopulation is invoked by calling *ConstructSubpopulation* (Algorithm 1, line 7). Algorithm 2 represents this concurrently-run constructor method that creates each subpopulation and processes them. Note that each subpopulation is processed on a separate thread. The array SP keeps a collection of references to each constructed subpopulation.

In ParallelClust, we employ a novel information sharing function. We experimentally compare our approach using 2 alternative information sharing functions, and also when no information sharing function is used.

Each isolated subpopulation enhances diversity, mitigating the potential for premature convergence to local optima [6]. Also note that each subpopulation is seeded differently so that the results of each subpopulation is different.

Every X generations (by default 3), each subpopulation shares information with other subpopulations. We experimentally found that sharing quite frequently (e.g. every 3 generations), produced the best results. Thus, our particular parameter setup of ParallelClust make it a fine-grained PGA, as it shares information quite frequently [6]. ParallelClust could also act as a more course-grained PGA if the *informationShareInterval* parameter (see Table 1) is set to a higher number, so that each subpopulation shares information less frequently.

Once all threads are complete, that is, all instances of *Construct Subpopulation* have run to completion, we finally get the highest fitness chromosome found among all subpopulations (Algorithm 1, line 10), returning the highest fitness chromosome, $CR_{globalbest}$.

Input : A dataset D_o, number of subpopulations S, number of
generations N, the information share interval X
Output: The best found chromosome $CR_{bestglobal}$

1 /* Store array of each subpopulation   */
2 Create array SP of size S;

3 /* Compo. 1: Normalize dataset   */
4 $D \leftarrow Normalize(D_o)$;

5 **for** $s \leftarrow 1$ *to* S **do**
6 | /* Run concurrent subpopulation algorithm   */
7 | $SP[s] \leftarrow$ ConstructSubpopulation(D, N, X);
8 **end**

9 /* Compo. 9- Best chromosome across all subpopulations   */
10 $CR_{globalbest} \leftarrow$ GetBestGlobal(SP);

11 **return** $CR_{globalbest}$

Algorithm 1. ParallelClust

Input : A dataset D, number of generations N, the information share interval X
Output: The best chromosome CR_{best}

1 /* Compo. 3: Initial population   */
2 $P_i \leftarrow$ GenerateInitialPopulation(D);
3 $P_s \leftarrow$ InitialSelection(P_i);

4 /* Store best chromosome for this subpopulation   */
5 $CR_{best} \leftarrow$ BestChromosome(P_s);

6 **for** $n \leftarrow 1$ **to** N *do* **do**
7 | **if** $n \bmod X \neq 0$ **then**
8 | | /* Compo. 8: Mutation   */
9 | | $P_c \leftarrow$ Crossover(P_s);
10 | **else**
11 | | /* Compo. 6: Information share every X generations   */
12 | | $P_c \leftarrow$ InformationShare(P_s);
13 | **end**
14 | /* Compo. 6: Elitism   */
15 | $CR_{best} \leftarrow$ Elitism(CR_{best}, P_c, D);
16 | /* Compo. 8: Mutation   */
17 | $P_c \leftarrow$ Mutation(P_c, D);
18 | /* Compo. 6: Elitism   */
19 | $CR_b \leftarrow$ Elitism(CR_b, P_c, D);
20 | **if** $g > 10$ **then**
21 | | /* Compo. 7: Chromosome Selection   */
22 | | $P_s \leftarrow$ SelectChromosome(P_s, P_c)
23 | **else**
24 | | $P_s \leftarrow P_c$
25 | **end**
26 **end**

27 /* Compo. 2: MK-Means (applied to best chromosome)   */
28 $CR_{best} \leftarrow$ MK-MEANS($CR_{best}, D, 50$);
29 $CR_{best} \leftarrow$ Denormalize(C_n, D, D_o);

Algorithm 2. Construct Subpopulation

Component 1: Normalize the Dataset
In Algorithm 1, the input is initial dataset of records, denoted by D_o, able to contain both numerical and/or categorical attributes.

All numerical attributes of the domain $A_j = [l, u]$, are normalized to the domain $[0, 1]$. This is done by $A_j = \frac{D_{i,j}^o - l}{u - l}$, where $D_{i,j}^o$ is the j-th attribute of the i-th record of the initial dataset D_o. A similar domain normalisation process is applied to categorical attributes.

Component 2: Modified K-Means (MK-Means)
We utilise MK-Means++, a modified K-Means++, which can handle both numerical and categorical attributes, as a hill-climber to improve the fitness of any given chromosome [3]. We initially to find a good set of chromosomes with a diverse range of viable k-values, running the MK-Means hill climber a given number of times to improve the solution further. Finally, after all generations of a subpopulation are complete (see line 29 of Algorithm 2), we apply MK-Means to our best found chromosome to improve our solution further, ensuring it is as close as possible to its nearest local optimum. Our chosen parameter values for how many times we run MK-means rely on those used by GenClust++ [3], which be seen in Table 1.

The distance (see below) between two records D_i and D_l is calculated as the average of all distances between each corresponding attributes of both records.

$$d(D_i, D_l) = \frac{\sum_{j=1}^{t} | D_{i,j} - D_{l,j} | + \sum_{j=t+1}^{m} d(D_{i,j}, D_{l,j})}{| A |} \quad (1)$$

Attributes are sequenced so the first t attributes are numerical, and subsequent $| A | - t = m - t$ attributes are categorical.

The distance among numerical and categorical attributes is computed using the Manhattan metric rather than the Euclidean metric.

Apart from these augmentations, MK-Means operates identically to K-Means [7].

Component 3: Initial Population and Initial Selection
We quickly determine a set of viable chromosomes with a variety of k-values so that the user does have to rely on domain or expert knowledge to know the number of clusters, that is, k. Thus, users do not require a priori knowledge regarding the dataset. Other parameters can simply use default values, or be easily set by the user to suit their particular needs.

A high-quality yet diverse initial population is likely to produce higher-quality clustering results [8].

To achieve a sound initial population we use the same approach of Gen-Clust++'s [3] Initial Population, defined as s, where s is the initial population size.

We initially create a base population of size $3 \times s$. This is done by generating 5 solutions for every k-value in the set $\{2, 3, 4, ...(3 \times s/10 + 1)\}$. Secondly, we produce 5 solutions for every random k-value in the range of $[2, \sqrt{| D |}]$. This provides a good spread of viable k-values to explore the solution space. The reason we repeat MK-Means is to improve the chances of producing the best results for that particular k-value, since the seed selection process has inherent randomness that can produce a diverse range of results.

We now filter down this initial population to a third of its original size, that is, from $3 \times s$ to s. GenClust++ uses a technique called probablistic selection [3], which involves finding the average for each k-value and choosing chromosomes one-by-one using the roulette wheel technique. This was not only convoluted in terms of the code, but is also quite computationally demanding, with no clear improvement to final clustering results. We instead use a simplistic method called *InitialSelection*, wherein we simply choose the best third of the initial population created. This yields the s number of chromosomes that form our initial population (or subpopulation). Our approach quickly and simply yields many high-quality candidates, while still garnering a variety of viable k-values.

Component 4: The Crossover Operation
As employed with GenClust++ [3], we first choose two parents to perform conventional one-point crossover. The first parent chosen is the highest fitness chromosome of a subpopulation, and then the second is chosen probabilistically from this subpopulation via the commonly used roulette wheel technique [9]. Then, an existing gene rearrangement operation [8] is applied to the second chromosome. This is to align the corresponding inferior chromosome's genes to the superior genes (improving the effectiveness of crossover operations as it reorders genes corresponding to those which are most similar). Now, the application of a conventional single point crossover can occur. This operation improves our genetic search by enhancing diversity.

Component 5: Periodic Information Sharing
A major advantage of having multiple subpopulations is the ability to information share and discover new viable solutions.

ParallelClust has the ability to allow a user-defined number for the periodic interval at which information sharing occurs, defined by the parameter *informationShareInterval*. We found that sharing quite frequently worked well on the datasets we experimented with. However, our approach offers users the flexibility to discover which setting works best for their particular use case and dataset.

With ParallelClust, we compare 3 different information sharing functions in isolation:

1. **replaceWorstChromosomeWithRandomBest (RB)**
   This is our novel technique that replaces the worst fitness chromosome of the subpopulation with a random subpopulation's best chromosome.
2. **replaceWorstChromosomeWithRandom (RR)**
   This is another alternative we propose that replaces the worst fitness chromosome of the subpopulation with a random chromosome from a random subpopulation.
3. **replaceWorstChromosomeWithBestFromNext2 (B2)**
   This is based on HeMI's [2] method of every 10 iterations replacing the worst fitness chromosome with only the next 2 neighboring subpopulation's best chromosome. This requires an ordered collection of subpopulations. For instance, subpopulation 1 would have neighboring subpopulations 2 and 3.

Component 6: The Elitism Operation
Based on GenClust [8], the core utility of the elitism operation is to ensure that the best performer for each population (subpopulation in the case of Parallel-Clust) exists within the population. This guarantees that the best chromosome found is not eradicated from a population, meaning that the final solution for that population can only improve upon the best solution found thus far. This is done by storing the best chromosome up until the current generation. If this has a higher fitness than the worst performing chromosome of the population (after crossover is performed), it replaces the worst performer. Finally, we also update the best performer of this subpopulation if we find any chromosomes with a higher fitness.

Component 7: Chromosome Selection
Since crossover profoundly alters chromosomes as a result of its inherent randomness, it can seriously degrade fitness of particular members of the population [3]. To avoid the severe loss of quality solutions within each subpopulation, this operation ensures that the best s chromosomes from the current and previous generation are retained, where s is the number of chromosomes in initial population. However, it is only applied by default from the 10th generation onward. This allows a wide exploration of the solution space during the early generations of the subpopulation.

Component 8: The Mutation Operation
Mutation introduces random exploration, adding the possibility to find slightly better solutions each generation [8].

The mutation probability P_j of the j-th chromosome CR_i in P_c is determined using this equation [3].

$$P_j = \begin{cases} \frac{f_{max} - fitness(CR_j)}{2(f_{max} - \overline{f})} & \text{when } fitness(CR_j) > \overline{f} \\ 1/2 & \text{when } fitness(CR_j) \leq \overline{f} \end{cases} \tag{2}$$

Note that $fitness(CR_j)$ is the fitness of the j-th chromosome in the current generation, and f_{max} is the fitness of the best chromosome in the current population, and $\overline{f}$ is the mean fitness of the all chromosomes in the current population.

In order to apply mutation to a chromosome, we randomly select a random gene and change a random attribute. The attribute is chosen uniformly across all attributes, and similarly the value is chosen uniformly within the relevant domain.

Component 9: Global Best Chromosome Selection
Similar to HeMI [2], after all the generations of a subpopulation have completed, we run MK-Means several times on the highest fitness chromosome to improve it further, and finally denormalise it. After all concurrently processed subpopulations have run to completion, we choose the best clustering solution across all subpopulations (Algorithm 1, line 10). We call this best solution $CR_{globalbest}$. Then, records are allocated to their nearest cluster centroids and the final clustering results are displayed.

Also note that we removed a component called Probabilistic Cloning with MK-Means from GenClust++ [3], as it consistently produced lower fitness solutions across the datasets experimented on with ParallelClust.

3 Results

Our experiments consisted of GenClust++ with one large population of 80 chromosomes, and ParallelClust (PC) with 8 subpopulations with 10 chromosomes per subpopulation. This enhances the comparability of both approaches as the total number of chromosomes are equal. Table 1 presents default parameter values for GenClust++ and ParallelClust (used in experiments). The non-italicised parameters of Table 1 are the GenClust++ [3] values ParallelClust uses.

Table 1. Parameter setup for GenClust++ and ParallelClust

	GenClust++	ParallelClust
Number of Generations	60	60
Number of Subpopulations	*N/A*	*8*
Initial Population Size	*80*	*10*
maxKMeansIterationsFinal	50	50
maxKMeansIterationsInitial	60	60
maxKMeansIterationsQuick	*15*	*N/A*
Seed	10	10
StartChromosomeSelectionGeneration	50	50
InformationShareInterval	*N/A*	*3*

The datasets (see Table 2) were downloaded as part of a collection of 37 classification problems available from the University of Waikato [10], originally sourced from UCI machine learning datasets [11]. Class attributes were not removed from the datasets, and were considered as regular attributes.

Table 2. Experimental Datasets

	No. records	Total attributes missing	No. numerical attributes	No. categorical attributes	No. Classes
Glass	214	0%	11	1	7
Vote	435	3.9%	0	17	2
Vehicle	946	0%	18	1	4
Vowel	990	0%	10	4	11
Segment	2310	0%	19	1	7

Tables 3 and 4 present experimental results showing cluster quality and runtime, respectively, for ParallelClust versus GenClust++. Each row presents a different a dataset, and each column is the particular algorithm used. We try our approach using ParallelClust's novel function replaceWorstChromosomeWithRandomBest (RB), and 2 alternative techniques, replaceWorstChromosomeWithRandom (RR) and replaceWorstChromosomeWithBestFromNext2 (B2). The best results for each particular dataset are shown in bold. The fitness function we used was the inverse of the Davies-Bouldin index (DBI^{-1}). Results are the average of 5 separate runs of each algorithm on that particular, as random seeding can produce slightly different results each run. Experiments were performed with a 2.5 GHz Intel Core i7, 4 core, dual-threaded machine with 16GB of RAM. The code was adapted from GenClust++ [3], and is available at https://github.com/StormBartlett/Clustering-Algorithm-ParallelClust.

Table 3. GenClust++ vs. ParallelClust (PC) Fitness (DBI^{-1})

	GenClust++	PC (No sharing)	PC (with RB)	PC (with RR)	PC (with B2)
Glass	**1.36**	1.33	1.34	1.34	1.31
Vote	0.55	0.55	**0.61**	0.59	0.60
Vehicle	0.93	0.945	**0.96**	0.94	0.96
Vowel	0.60	0.59	**0.65**	0.61	0.64
Segment	1.18	1.18	**1.2**	1.18	1.18

Table 4. GenClust++ vs. ParallelClust (PC) Runtime (ms)

	GenClust++	PC (No sharing)	PC (with RB)	PC (with RR)	PC (with B2)
Glass	9599	**3823**	6010	5850	6638
Vote	55433	**12832**	21032	22402	23563
Vehicle	84111	**28626**	38193	33461	36783
Vowel	96223	**32590**	57569	53366	57301
Segment	85948	**35234**	65023	72022	65400

In terms of cluster quality, 4 out of 5 datasets experimented with achieved a reasonable improvement in final cluster quality. The generally best performing information sharing function was found to RB, producing final clustering solutions with a roughly 1 to 10% improvement in fitness over GenClust++ [3].

ParallelClust's novel function, RB appears to perform the best with ParallelClust and versus GenClust++ [3] as it incorporates the idea of promoting diversity via randomly choosing a subpopulation, but still retains the concept of selection by choosing the most elite chromosome from that subpopulation. Perhaps function RR performed less effectively because by choosing a random subpopulation, and then a random chromosome, it overly promotes diversity, and generally does not exploit other high-quality solutions. The B2 [2] technique features less randomness, and enforces a more controlled dispersion of high-quality

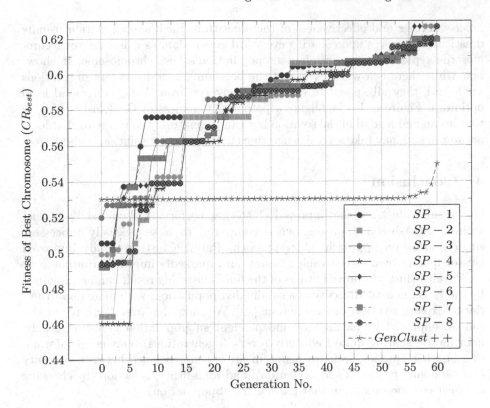

Fig. 4. ParallelClust (with RB) fitness of CR_{best} for each of ParallelClust's 8 subpopulations (SP) versus GenClust++'s single population over each generation (Color figure online)

solutions by only allowing the best solution to be taken from the next 2 neighboring subpopulations. As the best chromosome is always chosen from the same 2 subpopulations, this may diminish diversity. Further, RB and RR can enhance diversity as they both feature a stronger element of randomness in that a random subpopulation is chosen. ParallelClust with no information sharing function has consistently lower performance than ParallelClust, and sometimes below that of GenClust++ [3]. This indicates the importance of information sharing in maximizing the effectiveness of clustering with our PGA-based approach.

On all datasets, ParallelClust with no information sharing was the fastest as it avoids costly operations taken in RB, RR and B2 (e.g replacing the worst chromosome). However, we hold that the improved cluster quality offered by an information sharing function (e.g. RB) justifies the increase in processing time.

Figure 4 graphically presents each of ParallelClust's subpopulations versus GenClust++ [3] for one particular run on the Vote dataset. It shows how each subpopulation's best chromosome (CR_{best}) progressively improves over each generation. Being able to compare each subpopulation, and visually see the impacts when information sharing is performed, offers intuitive insight into the inner-workings and effectiveness of the algorithm. Each subpopulation usually drastically improves upon sharing every 3rd generation, as it has its worst chromosome replaced with a random subpopulation's best chromosome. It shows how while there is a wider spread at the beginning as a diverse set of solutions are found, they all appear to eventually converge around the best found local optimum. Figure 4 also highlights a potential weakness of GenClust++ [3], in that it can find a good initial solution but can prematurely converge to that local optimum, and may take several generations before a better solution is discovered.

4   Conclusion

Our results illustrate the improved clustering performance offered by a fine-grained PGA-based approach, when compared to a sequentially processed, single-population approach. Our approach, ParallelClust, generally improved cluster quality by a considerable amount, and markedly improved runtime.

ParallelClust's improved fitness of the final clustering result can be explained by the larger genetic diversity of a subdivided population, with more opportunities to explore a variety of local optima [12]. We show that overall cluster results can be improved by having each independent subpopulation perform periodic information sharing quite frequently (every 3 generations). Our novel information sharing function seeks to maintain an optimal balance between diversity (via randomly picking a subpopulation), and high-fitness selection (by choosing the best chromosome from that particular subpopulation).

References

1. Hendricks, D., Gebbie, T., Wilcox, D.: High-speed detection of emergent market clustering via an unsupervised parallel genetic algorithm. South Afr. J. Sci. **112**, 57 (2016)
2. Beg, A.H., Islam, Md.Z., Estivill-Castro, V.: Genetic algorithm with healthy population and multiple streams sharing information for clustering. Knowl.-Based Syst. **114**, 61–78 (2016)
3. Islam, Md.Z., Estivill-Castro, V., Rahman, Md.A., Bossomaier, T.: Combining k-means and a genetic algorithm through a novel arrangement of genetic operators for high quality clustering. Expert Syst. Appl. **91**, 402–417 (2018)
4. Cavuoti, S., Garofalo, M., Brescia, M., Pescape', A., Longo, G., Ventre, G.: Genetic algorithm modeling with GPU parallel computing technology. In: Apolloni, B., Bassis, S., Esposito, A., Morabito, F. (eds.) Neural Nets and Surroundings. Smart Innovation, Systems and Technologies, vol. 19, pp. 29–39. Springer, Heidelberg (2013). https://doi.org/10.1007/978-3-642-35467-0_4
5. Davies, D.L., Bouldin, D.W.: A cluster separation measure. IEEE Trans. Pattern Anal. Mach. Intell. PAMI **1**(2), 224–227 (1979)
6. Li, X., Kirley, M.: The effects of varying population density in a fine-grained parallel genetic algorithm, vol. 2, pp. 1709–1714, February 2002
7. Tan, P.N., Steinbach, M., Kumar, V.: Introduction to Data Mining. Pearson Addison Wessley, Boston (2005)

8. Rahman, Md.A., Islam, Md.Z.: A hybrid clustering technique combining a novel genetic algorithm with k-means. Knowl.-Based Syst. **71**, 21–28 (2014)
9. Maulik, U., Bandyopadhyay, S.: Genetic algorithm-based clustering technique. Pattern Recogn. **33**, 1455–1465 (2000)
10. University of Waikato - collections of datasets. https://www.cs.waikato.ac.nz/ml/weka/datasets.html. Accessed 7 July 2018
11. Frank, A., Asuncion, A.: UCI machine learning repository (2010). http://archive.ics.uci.edu/ml. Accessed 7 July 2018
12. Kohlmorgen, U., Schmeck, H., Haase, K.: Experiences with fine-grained parallel genetic algorithms. Ann. Oper. Res.-Ann. OR **90**, 203–219 (1999)

A Learning Approach for Ill-Posed Optimisation Problems

Jörg Frochte[1(✉)] and Stephen Marsland[2]

[1] Bochum University of Applied Sciences, 42579 Heiligenhaus, Germany
joerg.frochte@hs-bochum.de
[2] Victoria University of Wellington, Wellington, New Zealand
stephen.marsland@vuw.ac.nz

Abstract. Supervised learning can be thought of as finding a mapping between spaces of input and output vectors. In the case that the function to be learned is multi-valued (so that there are several correct output values for a given input) the problem becomes ill-posed, and many standard methods fail to find good solutions. However, optimisation problems based on multi-valued functions are relatively common. They include reverse robot kinematics, and the research field of AutoML – which is becoming increasingly popular – where one seeks to establish optimal hyperparameters for a learning algorithm for a particular problem based on loss function values for trained networks, or to reuse training from previous networks. We present an analysis of this problem, together with an approach based on k-nearest neighbours, which we demonstrate on a set of simple examples, including two application areas of interest.

Keywords: Multi-valued functions · Ill-posed optimisation · Local models · AutoML

1 Introduction

Consider a standard regression problem. Given a training set $S = (x, y)$ of pairs of input and target data, we aim to identify a function $h(x)$ such that

$$y = h(x) + \varepsilon, \tag{1}$$

where the ε term gives the residual error and is a function of the data and the particular form of regression used.

Implicit in this model is the assumption that the problem is well-posed, i.e.,

1. a solution exists,
2. the solution is unique, and
3. the solution is stable (its behaviour changes continuously with respect to the initial conditions).

Ill-posed problems – a class that includes many inverse problems – can be very important. We will focus on problems that do not have a unique solution, so that there is more than one possible target for a given input, which are known as multi-valued functions. As shall be discussed shortly, this class of problems includes inverse kinematics and a form of automated machine learning (AutoML).

Standard regression techniques do particularly poorly on these problems, finding solutions between the two possible correct results. For example, consider the cylindrical spiral given by $y^2 = \tan^{-1}\left(\frac{x_2}{x_1}\right)$, $x_1, x_2 \in [-1, 1]$, which is a variant on that suggested in [9]. Note that this corresponds to the mapping between Cartesian and polar coordinates. As can be seen on the left of Fig. 1, there are two correct targets for each input pair (x_1, x_2), and the multilayer Perceptron (MLP) fails to find either solution. Local approaches such as k-nearest neighbours also fail, albeit in a slightly different way. We will revisit this problem in Sect. 3.1, but on the right of Fig. 1 demonstrate that our approach finds both solutions to this multi-valued problem.

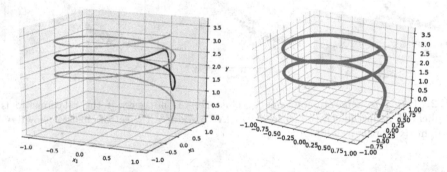

Fig. 1. *Left:* The MLP (blue) finds an incorrect solution to a multi-valued function consisting of the spiral shown in red. *Right:* Our approach finds both solutions. (Color figure online)

The heart of the problem lies in the fact that so-called multi-valued functions define a left-total binary relationship [8]. A binary relationship takes ordered pairs $(x \in X, y \in Y)$ for sets X and Y, and assigns it to some subset of the Cartesian product $X \times Y = \{(x, y) : x \in X, y \in Y\}$. As it is defined on ordered pairs, there can be many y that match an x and vice-versa, and there may also be some $y \in Y$ that do no have a relationship with any x and vice versa; see also [5], which considers the learning of such relationships. A left-total binary relationship requires that at least one y exists for each x, but does not require that it is unique. Obviously, such functions are not globally invertible.

This is why inverse problems so commonly have this form: consider the inverse kinematics problem from robotics. In this classic regression problem, the old adage that all roads lead to Rome (i.e., there are many paths to reach any given location of the end-effector) means that the inverse is not properly defined. While adding regularisation functions such as aiming to minimise energy or time can help, there is still no guarantee of a unique, and therefore potentially invertible, solution.

One way to deal with this lack of global invertibility is to appeal to the implicit function theorem, and create local invertible approximations of the function. Providing that the regions of the domain with matching points in the codomain are sufficiently separated, this can work well, as was shown in [6] where networks of radial basis functions were used to approximate the multi-valued function. However, there will always by pathological examples where this assumption breaks down. As was discussed in [6], there is no guarantee that an appropriate partition of the data can be found, and even if there is, the implicit function theorem requires that the function is continuously differentiable, which may not be true in general.

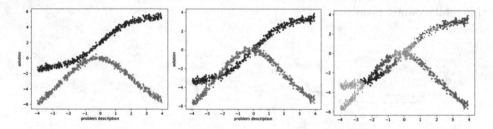

Fig. 2. Trying to separate solutions by clustering is not easy if the relationship is complicated, as in the centre picture. DBSCAN produces the set of clusters shown on the right. (Color figure online)

In fact, this problem can be made worse if the data is clustered first to aid in the selection of partitions, as Fig. 2 shows. If both solutions (red and black) are easy to separate, as in the left problem set, clustering works well. If their relationship is more complicated, as in the middle of the figure, one ends up with a lot of problems. The right plot shows the result of using DBSCAN [4], which is dependant on the parameters. As one can see, the subsets identified by clustering do not necessarily consist of data from one solution (which we call 'pure'). Increasing the number of clusters makes them pure, but tends towards local learners.

For some problems there is additional structure to the problem that can be exploited. This is true for inverse kinematics, where there is a time progression that gives order to the data samples: the data is generated as a sequence $f_i : t \mapsto \mathbb{R}^n = y$ of several functions f_i, each providing a mapping between the same start and end points. The set of different trajectories performing the same mapping provide enough structure for neural networks to find good approximations, such as the RNNs used by [13] and the standard feedforward networks of [9] and [1]. Beyond this, [11] presents an approach using regularisation networks that includes learning an algebraic representation of the multi-valued function, while in [2] the authors extend a hierarchical Dirichlet process hidden Markov model to a multi-valued function regression. One of the most recent publications in this branch is [3], which uses an approach based on an infinite mixture of linear experts, thus enabling online learning.

Consider now the example of automatic machine learning (AutoML), which aims to automate the whole process of selecting machine learning algorithms, fitting hyperparameters, and optimising them, see e.g., [14]. One way to formulate this problem is as an optimisation problem. Data are presented as triples consisting of a description of a learner such as a neural network, a set of weight values for it, and the value of the loss function of that network on some dataset: $(\theta, x, f_D(x, \theta))$, where the D subscript labels the dataset that was used for testing. A set of such triples comprise the training set for an optimisation problem. In more general terms, this can be written as $\min_x f(x, \theta)$, where $f(\cdot)$ is the objective function (e.g., the neural network loss function, or some function to minimise such as energy, or cost), x are the variables we wish to optimise over (the weights of the neural network), and θ is a set of parameters that specify the precise problem. Note that this formulation includes multi-objective optimisation automatically using some of the elements of vector θ as weights for these parts of the objective function, e.g., $f(x, \theta) = \theta_1 f_1(x, \theta) + \theta_2 f_2(x, \theta)$.

In general, the dataset consists of noisy samples, such as the weights and loss function values of trained neural networks with particular sets of hidden layers, or certain examples of inverse kinematic solutions. In neither case is it likely that the dataset contains the actual global optimum.

We now present an extension to the k-nearest neighbour algorithm for optimisation of multi-valued functions, and demonstrate that it is well-behaved with respect to hyperparameter choices on a variety of test cases, including a simple example from AutoML.

2 A Clustering-Based Algorithm for Multi-valued Functions

Our algorithm is a variant of the partition approach that was described previously. For a given value of vector θ, we identify the points in the dataset that are closest (in the Euclidean norm) to that value and then cluster those points into c or fewer clusters based on a weighted sum of their coordinates using kNN (where $d_i = \theta - \theta_i$ and smear is a parameter that smoothes the kNN regression):

$$x(\theta) = \sum_{i=1}^{k} \omega_i x_i \text{ with } \omega_i = \frac{(d_i + \frac{\text{smear}}{k})^{-1}}{d} \text{ and } d = \sum_{i=1}^{k} \left(d_i + \frac{\text{smear}}{k}\right)^{-1} \quad (2)$$

The pseudocode for our algorithm is given in Algorithm 1. The algorithm works in three stages: we find a large set of points close to θ, and then refine it to the points in that set with smallest $f(\theta, x)$ values. These points are then

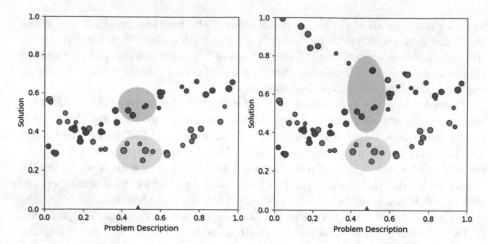

Fig. 3. It is not necessary to identify the number of clusters precisely. The algorithm is searching for points close to the θ value marked with a black triangle. *Left:* When there are two clusters, setting $c = 2$ means that two clusters are identified. However, *right:* the same parameter value when there are three clusters will merge two of the clusters, resulting in a spread of potential points. However, when the points nearest to each other in every cluster are identified, only the blue dots in the blue cluster will be considered. (Color figure online)

clustered into c clusters, which are post-processed to remove clusters with fewer than k members, and retain the k points that are nearest with respect to θ for the rest. It is not necessary to specify c exactly, as is shown in Fig. 3.

The k samples in a cluster are used with (2) to find the regression value of $f(x, \theta)$. The values at $f(x_i, \theta)$ could be used as weights, but we found that this did not improve the results much, and was computationally more expensive.

In the clustering step it is possible to use any clustering algorithm. We have tested k-means and fuzzy C-means, as they have a small number of hyperparameters and distribute points across as many clusters as possible. Note that although kNN is often described as a lazy learner, it is common to build a kd-tree or similar of all the distances to enable nearest neighbours to be found as efficiently as possibly. This is the approach used in e.g., scikit-learn [10].

Finally, we note that although the algorithm is intended to work with the function values $f(x, \theta)$ (score), it is possible to use it in cases where this information is not available. In that case $n = 1$, and there is no need to perform line 3 and the regression of f at the final step in line 7. The first example in the next section considers this type of example.

3   Experiments and Analysis

3.1   Regression Without Score

We first consider the example shown in the Introduction, in order to show that the algorithm is equivalent to other multi-valued function learners if the function

---

Algorithm 1. The kNN-MV Algorithm

1: **procedure** KNN-MV$(x, \theta, f(x, \theta))$

Require: $k \geq 1, c \geq 1$ ▷ $k =$ #neighbours, $c =$ #output classes

Require: $2 \leq m \leq 4, 1 \leq n \leq 3$ ▷ m and n control #points in neighbourhoods

2:      $\hat{N} =$ the indices of the $a \cdot k \cdot m \cdot n$ nearest neighbours of θ

3:      $N =$ the $a \cdot k \cdot m$ examples in $\hat{N}$ with lowest values of $f(x, \theta)$

4:      cluster the points in N into c clusters according to distance with respect to θ ▷
     e.g., using k-means

5:      **for** each cluster **do**

6:         **if** cluster has $< k$ members **then** return $\emptyset$

7:         **else** use the k points closest to each other for the regression of θ and $f(x, \theta)$
     using (2)

8:         **end if**

9:      **end for**

10: **end procedure**

---

values $f(x, \theta)$ are not provided. We sampled 20,000 points from the curve shown in red in Fig. 1, adding noise at the boundaries 0 and $\sqrt{4\pi}$, as they tend to overlap, leading to more choices, and set $c = 2$.

Table 1 shows that the algorithm works well and is stable with respect to added noise; see also the right of Fig. 1. It is hard to compare exactly with [9], as not all details are provided there, but they report a root mean square error (RMSE) of about $1.85 \cdot 10^{-2}$, which is comparable to ours, and we make fewer assumptions about the structure of the solution.

Table 1. Results using the parameters $k = 3$, smear $= 1$, $a = 2$, $m = 4$.

	Percentage of noise on x in training data				
	0%	1%	2%	3%	4%
One correct answer	100 %	100 %	100 %	100%	100%
Two correct answers	91.2 %	92.8%	91.0%	91.9 %	89.9%
AME	$6.5 \cdot 10^{-4}$	$8.2 \cdot 10^{-3}$	$1.6 \cdot 10^{-2}$	$2.5 \cdot 10^{-2}$	$3.2 \cdot 10^{-2}$
RMSE	$8.1 \cdot 10^{-4}$	$1.1 \cdot 10^{-2}$	$2.2 \cdot 10^{-2}$	$3.4 \cdot 10^{-2}$	$4.4 \cdot 10^{-2}$

3.2 An Analytical Test Case

As a simple example of the core usage of our method, we considered the function:

$$f(x, \theta) = 3 - \exp(20(-(x_1 - \theta_1)^2 - (x_2 - \theta_2)^2))$$
$$- \exp(20(-(x_1 - \theta_2 + 1)^2 - (x_2 - 0.5\theta_1^2)^2))$$
$$- 0.95 \exp(20(-(x_1^2 - \theta_2) - (x_2 + \theta_3 + 0.5)^2))$$

The problem consists of finding the multiple solutions to $\arg\min_{x\in\mathbb{R}^2} f(x,\theta)$ for fixed $\theta \in \mathbb{R}^3$.

We created 429,618 samples by computing numerical solutions by gradient descent from random initial starting points. Note that the space also has local minima, and so many of these numerical solutions will have become stuck in them. Table 2 shows the results with $k = 3$ and $k = 5$ for both the standard kNN and our multi-value version, while Fig. 4 looks at the effect of the two parameters, m and n. The table shows that the method is far more successful that standard kNN, even at finding one solution, while the figure demonstrates that the algorithm is not very sensitive to the choice of the parameters. In general, $1.5 \le n \le 2.5$ and $2 \le m \le 4$ is a good range. Beyond this one can see in Table 2 that using the score in kNN-MV, which means in this application the value of the function f, leads to better results.

Table 2. Results using kNN and kNN-MV ($k = 3$ $k = 5$, $m = n = 2.5$).

Method	% answers with		Average error
	One result	Both results	on answers
Standard kNN ($k = 3$)	21.70%	00.00%	0.0042
kNN4MV ($k = 3$) without score	89.44%	62.12%	0.0072
kNN4MV ($k = 3$) with score	99.82%	83.20%	0.0083
Standard kNN ($k = 5$)	13.11%	00.00%	0.0357
kNN4MV ($k = 5$) without score	91.85%	64.48%	0.0073
kNN4MV ($k = 5$) with score	99.92%	85.35%	0.0094

3.3 Shot on Goal Learning

We now present a more interesting example, which combines kinematics – as common application area – with an ill-posed optimisation problem. Suppose that a robot is aiming to kick a ball into a soccer goal in such a way that it goes over the goalkeeper's head, as shown on the left of Fig. 5, by learning from examples. We model the path of the ball via a non-linear ordinary differential equation that includes the four fixed components of θ shown on the right of the figure with their allowable ranges. In fact, we will allow the wind speed θ_w to vary, and assume that the agent knows only the sign of it, not the value. The controllable parameter is the velocity of the ball $x(t)$ at $t = 0$. We represented $x(0)$ in polar coordinates as (φ, s) pairs of direction and speed.

We judged that a goal was scored if the ball arrives at the goal at a height between 2.1 and 2.4 m above the ground, with angle of travel $\varphi < -0.5°$. We created a large number of examples that satisfied these criteria, and gave them a score based on their height and angle when the ball crossed the goalline: $f(x,\theta) = x_2 + |\varphi|/2$, i.e., height + half the angle.

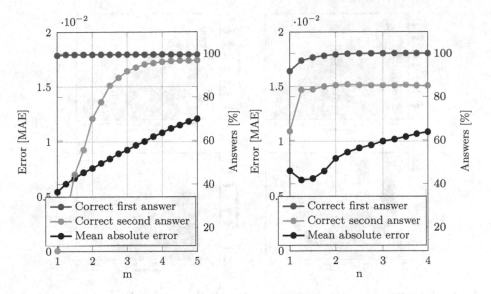

Fig. 4. Results of varying m with a constant $n = 2$ (*left*) and varying n for a constant $m = 2.5$ (*right*).

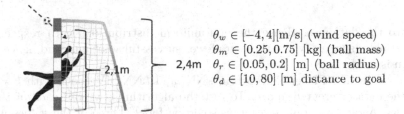

$\theta_w \in [-4, 4][m/s]$ (wind speed)
$\theta_m \in [0.25, 0.75]$ [kg] (ball mass)
$\theta_r \in [0.05, 0.2]$ [m] (ball radius)
$\theta_d \in [10, 80]$ [m] distance to goal

Fig. 5. Schematic showing a successful shot on goal, and the parameters of θ with their allowable range.

There are two very different strategies that can achieve this successfully, corresponding to two different angles of initial velocity. The higher angle strategy receives higher scores, but because the ball is in the air longer, it is more affected by the random wind. We computed 30,000 successful goal scoring examples by brute force as an initial training set. Our approach to this was as follows: (φ, s) pairs and values for θ were chosen by kNN-MV and evaluated. Those that were successful in scoring a goal were saved, with their score. For the others, we performed a search around that φ of $\pm 5°$, reducing that to $\pm 1°$ as more entries were added to the dataset, and finally stopping using the search at all.

The top line of Fig. 6 shows the final distribution of samples in the training set for two parameters of θ, ball radius and distance to the goal, while the bottom line shows them with respect to the two components of x, φ and s. Samples were created uniformly at random, but only successful samples were

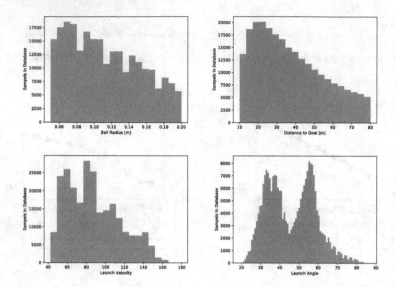

Fig. 6. Histogram of the distribution of solutions in the training database with respect to *top row:* fixed parameters (for a datapoint) θ_r and θ_d and *bottom row:* solution parameters φ and s.

added to the database. Hence the non-uniform distributions with respect to these variables suggest where there are fewer successful solutions, and hence the problem is harder.

We train a multi-layer Perceptron, kNN, and kNN-MV (both with $k = 5$) using the dataset created above. To test the algorithms, we ran each of them as follows. A random configuration was chosen for θ. Three of the values were then held fixed, and four samples of wind strength θ_w were chosen (which the learner does not know), all with the same sign (which the learner does know). The learner than generated four choices of (φ, s). The percentage of successful shots, and the score achieved by the most successful shot, are given in Table 3. Note that this is not an easy problem, and we consider that the near 40% of kNN-MV is a good result.

Table 3. Results of learning agent using three supervised methods for different limits for the distance to the goal.

Method	Maximum Distance									
	20 m		30 m		40 m		60 m		80 m	
	%	Score	%	Score	%	Score	%	Score	%	Score
MLP	4.74	2.71	0.21	2.70	0.60	2.70	0.67	2.68	1.65	2.68
kNN	15.11	2.70	11.83	2.68	9.81	2.66	8.35	2.65	6.88	2.64
kNN-MV	36.82	2.60	39.14	2.59	37.12	2.59	33.20	2.59	27.30	2.58

3.4 A Simple AutoML Example

The fact that a multi-layer Perceptron with a single hidden layer of 2 nodes and a total of 9 weights can solve the XOR problem is well-known. There are actually six different solutions for the weights, up to scaling, which arises because there are three different ways to construct XOR from more basic logical operators:

$$x_1 \text{ XOR } x_2 = (x_1 \wedge \neg x_2) \vee (\neg x_1 \wedge x_2) \tag{3}$$
$$= (x_1 \vee x_2) \wedge (\neg x_1 \vee \neg x_2) \tag{4}$$
$$= (x_1 \vee x_2) \wedge \neg (x_1 \wedge x_2) \tag{5}$$

The first bracket term is represented by one hidden unit and the second by another, while the final AND/OR-operation is performed by the output layer. Hence, the network has three ways to choose the weights, and a symmetry in the order of the operators.

One goal of AutoML is to reuse old models to speed-up the design of new ones. Learning the weights of a neural network involves solving a multi-valued problem because different configurations of the weights can result in very similar performances, see e.g. [7,12].

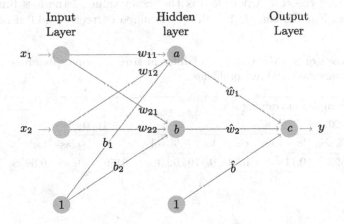

Fig. 7. MLP with one hidden layer for XOR

We took a standard XOR problem, and applied a rotation $\alpha \in [-45°, +45°]$ and a translation $[t_1, t_2]$, with $0 \leq t_i \leq 0.5$. The result is the set illustrated in Fig. 8.

Our AutoML problem is to find values for the 9 weights of the neural network (see Fig. 7) based on a set of trained networks for different values of $\theta = (\alpha, t_1, t_2)$ with the assistance of values $f(x, \theta)$ being the value of the loss function (sum-of-squares error). The database only contains solutions with $f(x, \theta) < 0.1$.

Table 4 shows the results on this example problem for training databases with 500 and 1000 samples in the training set. In each case kNN-MV was able

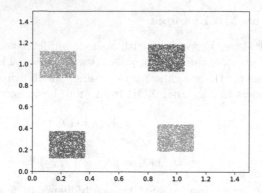

Fig. 8. Affine transformation of an XOR Training Set.

to find at least one solution. The algorithm produced 600 solutions during the test, since there are 6 possible different weights (up to scaling). *% sol.* is the percentage of those solutions that were found. As mentioned before, kNN-MV can predict the quality of the solutions as well. Therefore the column $|q - \text{loss}|$ is the mean difference between the predicted value of the loss function and the real one during the test, where loss is the mean value of the loss function for that solution. Finally, *class* is 1 if all are classified correctly and 0 if none are.

Table 4. Success of KNN-MV ($k = 5$) of different sizes of training databases, all values are mean average over 100 test problems.

Samples in training set							
500				1000			
% sol.	$|q - \text{loss}|$	class	loss	% sol.	$|q - \text{loss}|$	class	loss
89.6%	0.11	0.94	0.170	90.3%	0.0479	0.98	0.0998

4   Conclusions

In this paper we have considered the setting of multi-valued functions in general, and then presented a kNN variant that learns about these functions in general, and for ill-posed optimisation application cases in particular. Our algorithm shows good results on a set of example problems, which includes a simple application in AutoML. We have shown that a global approach decomposing the multi-value functions cannot be performed without making strong assumptions concerning the nature of the database, e.g. time-series data. Nevertheless, one of our future prospects is to use the presented method as the nucleus of a mixed lazy and eager learner with the goal of achieving higher-order regression on trusted subsets of the database. We will also be applying our algorithm to real-world, higher-dimensional datasets. The used datasets and simulation codes in this paper are published on the authors website.

References

1. Brouwer, R.K.: Feed-forward neural network for one-to-many mappings using fuzzy sets. Neurocomputing **57**, 345–360 (2004)
2. Butterfield, J., Osentoski, S., Jay, G., Jenkins, O.C.: Learning from demonstration using a multi-valued function regressor for time-series data. In: 2010 10th IEEE-RAS International Conference on Humanoid Robots (Humanoids), pp. 328–333. IEEE (2010)
3. Damas, B., Santos-Victor, J.: Online learning of single-and multivalued functions with an infinite mixture of linear experts. Neural Comput. **25**(11), 3044–3091 (2013)
4. Ester, M., Kriegel, H.P., Sander, J., Xu, X.: A density-based algorithm for discovering clusters in large spatial databases with noise. In: Proceedings of the Second International Conference on Knowledge Discovery and Data Mining, pp. 226–231 (1996)
5. Goldman, S.A., Rivest, R.L., Schapire, R.E.: Learning binary relations and total orders. SIAM J. Comput. **22**(5), 1006–1034 (1993)
6. Hahn, K., Waschulzik, T.: On the use of local RBF networks to approximate multivalued functions and relations. In: Niklasson, L., Bodén, M., Ziemke, T. (eds.) ICANN 1998. PNC, pp. 505–510. Springer, London (1998). https://doi.org/10.1007/978-1-4471-1599-1_75
7. Hecht-Nielsen, R.: Theory of the backpropagation neural network. In: Neural Networks for Perception, pp. 65–93. Elsevier (1992)
8. Kilp, M., Knauer, U., Mikhalev, A.: Monoids, Acts and Categories, With Applications to Wreath Products and Graphs. de Gruyter (2000)
9. Lee, K.W., Lee, T.: Design of neural networks for multi-value regression. In: 2001 Proceedings of International Joint Conference on Neural Networks, IJCNN 2001, vol. 1, pp. 93–98. IEEE (2001)
10. Pedregosa, F., et al.: Scikit-learn: machine learning in Python. J. Mach. Learn. Res. **12**, 2825–2830 (2011)
11. Shizawa, M.: Multivalued regularization network-a theory of multilayer networks for learning many-to-h mappings. Electron. Commun. Jpn. (Part III: Fundam. Electron. Sci.) **79**(9), 98–113 (1996)
12. Sussmann, H.J.: Uniqueness of the weights for minimal feedforward nets with a given input-output map. Neural Netw. **5**(4), 589–593 (1992)
13. Tomikawa, Y., Nakayama, K.: Approximating many valued mappings using a recurrent neural network. In: Proceedings of the 1998 IEEE International Joint Conference on Neural Networks, IEEE World Congress on Computational Intelligence, vol. 2, pp. 1494–1497. IEEE (1998)
14. Wong, C., Houlsby, N., Lu, Y., Gesmundo, A.: Transfer learning with neural AutoML. In: Advances in Neural Information Processing Systems, pp. 8356–8365 (2018)

Topic Representation using Semantic-Based Patterns

Dakshi Kapugama Geeganage[✉], Yue Xu, and Yuefeng Li

Queensland University of Technology, Brisbane, Australia
dakshi.geeganage@hdr.qut.edu.au,
{yue.xu, y2.li}@qut.edu.au

Abstract. Topic modelling is the state of the art technique for understanding, organizing, and extracting information from text collections. Traditional topic modeling approaches apply probabilistic techniques to generate the list of topics from collections. Nevertheless, human understands, summarizes and discovers the topics based on the meaning of the content. Hence, the quality of the topic models can be improved by grasping the meaning from the content. In this paper, we propose an approach to identify sets of meaningful terms based on ontology, called Semantic-based Patterns, which represent the content of a collection of documents. A set of related semantic-based patterns can be used to represent a latent topic in the collection. The proposed Topic Representation using Semantic-based Patterns aims to generate semantically meaningful patterns based on ontology rather than term co-occurrence as what existing topic modelling methods do. The semantically meaningful patterns were evaluated by applying the information filtering to semantic-based topic representation. The semantic based patterns were used as features for information filtering and were evaluated by comparing against popular information filtering baseline systems. Topic quality was evaluated in terms of topic coherence and perplexity. The experimental results verified that the quality of the proposed patterns was better than features used in baseline systems for information filtering. Further, the quality of topic representation outperforms the generated topics of other topic modeling approaches.

Keywords: Patterns · Semantics · Topic representation · Concepts

1 Introduction

Today, we are living in a world where digital information surrounding us has been an essential element. People are too much relying on electronic text than ever before with the interactions of "web, social media, instant messaging to online transactions, government intelligence, and digitized libraries" [1]. Variety, complexity and the volume of content demoralize the possibility of human intervention for information analysis, organization and searching. Different techniques such as keywords, phrases, patterns and domain specific sentiments have been applied to enhance the understanding to text content. Topic modelling has become the most popular technique and the state of the art for understanding and summarizing information in a collection of documents. Existing topic models apply the probabilistic approaches [2–4] to derive the topics from content. Accordingly, word frequency and co-occurrences are considered to generate the topics. Word2Vec and doc2vec [5, 6] are two popular approaches to capture the

© Springer Nature Singapore Pte Ltd. 2019
T. D. Le et al. (Eds.): AusDM 2019, CCIS 1127, pp. 28–40, 2019.
https://doi.org/10.1007/978-981-15-1699-3_3

semantics in content based on deep learning. Nevertheless, the generated representation to understand the latent semantics is not interpretable.

Main contribution of this research is to introduce an interpretable technique to capture the meanings of content during the topic modeling process with the semantic perception. We use an external general knowledge base to discover the meanings based on content of the collection. Latent Dirichlet Allocation (LDA) [2] is the most popular and stable topic model among the probabilistic models. Ignoring the semantics of a given content can be considered as the main weakness of LDA. Patterns-based approaches, semantic based approaches using ontology and external knowledge sources were introduced to overcome the limitations of existing topic models [7–11]. But most of them are suffered from the problems related to accuracy and the quality of topics. Nevertheless, none of the approaches had successfully addressed the main problem associated with the LDA topic model. Therefore, it is important to introduce a topic model, which understands the meaning of text content and generate most appropriate set of topics.

A pattern can be defined as an association of terms and patterns are more meaningful than individual terms. Usually these patterns are generated based on frequency which is a statistical feature of patterns, but not generated based on semantic information. In this paper, we propose an approach, called Topic Representation using Semantic based Patterns (TRuSP), to identify sets of meaningful terms called Semantic-based Patterns based on ontology to represent the content of a collection of documents. The proposed approach, TRuSP generates the semantic-based patterns from a given collection by understanding the semantics of the content. Two algorithms are proposed to grasp the semantics in the documents in an effective way. The first algorithm groups related terms together based on their semantic similarity and relevance. The groups of terms are called as semantic cliques. Then the second proposed algorithm further categorizes the terms in each semantic clique into patterns based on the matching concepts in an ontology. In this paper, WordNet [12] is used as the lexical knowledge base to generate the semantic cliques by grouping the related terms together. The ontology, Probase [13], is used as the concept knowledge base to generate semantic patterns. Semantic patterns are generated by grouping terms in each clique to interpret the meaning of the semantic clique. The semantic-based patterns are created based on the conceptualization of terms. The semantic cliques and semantic-based patterns can be defined as the meaningful elements to represent a document collection. Further, each semantic clique can be considered as one topic in the collection and semantic-based patterns in the semantic clique can be considered as meaningful constructs to represent the meaning of the topic.

Section 2 of the research paper describes the research efforts related to semantic and pattern-based approaches in topic modeling. The proposed approach is explained in Sect. 3. Section 4 elaborates the information filtering based and topic quality-based evaluation with the experimental results of baseline systems. Section 5 provides the discussion about the results and finally, Sect. 6 concludes the paper.

2 State of the Art

Probabilistic, semantic based, pattern-based and hybrid approaches can be defined as the major research efforts conducted under the topic-modeling domain. Probabilistic Latent Semantic Analysis (PLSA) [4] was introduced to overcome the limitations of LSA [14].

Latent Dirichlet Allocation (LDA) [2] was introduced as an enhanced probabilistic topic model to overcome the limitations of LSA and PLSA. LDA is based on a three-level hierarchical Bayesian model, which captures the content of the collection as random mixtures over latent topics. Many topic modeling approaches have been introduced as extensions to the LDA. Topical N-Gram [15] considers the phrases during the topic generation process and relies on the semantic co-occurrences of words with phrases. "Maximum Matched Pattern-based Topic Model (MMPBTM)" [8] was developed to generate a pattern-based topic model on collections of documents. However, it is challenging to correctly identify the meaningful patterns from unstructured text content.

Some researches realized the importance of the semantics and embedded the knowledge into traditional topic models. Hence, external knowledge bases have been integrated with LDA. Probase [13] is a taxonomy that understands the natural language terms by grasping the knowledge from content of large amount of web pages. Yao et al. [9] and Tang et al. [10] used Probase as the external knowledge base and introduced semantic based topic models. CLDA [10] is a four-layer topic model which contains an intermediate concept layer to existing LDA model. Probase-LDA [9] is also a combination of LDA and Probase ontology. Yao et al. [9] applied k-medoids clustering to cluster the identified concepts and computed asymmetric dirichlet priors based on that. Probase-LDA [9] outperformed DF-LDA [7] and GK-LDA [16]. In most of the approaches, concepts have been extracted for single words to interpret the meanings and concept values have been incorporated with LDA topic generation. It is indeed challenging to apply pattern mining and semantic capturing techniques in unstructured text content.

3 Proposed Approach

Similar to topic modelling techniques, the goal of the proposed TRuSP is to generate a representation to represent the content of a given document collection. However, TRuSP is a novel technique which is very different from traditional topic modelling such as LDA in the sense that TRuSP is a semantic based topic representation approach which is based on both statistical features and semantic information rather than only statistical features such as term frequency and co-occurrences. TPuSP contains two major components to identify the semantic elements from a collection of documents. The first component takes account of both statistical and semantical information of terms to group frequent terms into cliques based on the relevancy and semantic similarity of the terms. We consider each clique may indicate a potential topic in the collection. The second component discovers semantically meaningful patterns within each clique based on the matching concepts in ontology.

3.1 Identifying Cliques of Related Terms

Aim of the first component is to identify the semantically meaningful cliques of terms from a collection of documents. Algorithm 1 identifies the cliques of semantically related terms from the document collection D. Let D be the collection of documents and T_D be a set of unique terms after the pre-processing. First, document frequency-based filtering is applied to filter terms t into T_D. For each term in T_D, its document frequency (DF) value should be larger than a threshold value (0.15 was taken in our experiments). WordNet [12] is used as the lexical knowledge base to group semantically related terms in T_D. In WordNet, each word is semantically well defined by the description, synonyms and the entailment words. In this paper, for each term in T_D, we propose to use its description, synonyms and entailment words to represent the term. For a term, let $des(t), syn(t), ent(t)$ denote a set of words in the description, synonyms, and entailment respectively. $Des(T_D), Syn(T_D), Ent(T_D)$ are the description words, synonyms, and entailment words for all the terms in T_D, which are defined as $Des(T_D) = \bigcup_{t \in T_D} des(t), Syn(T_D) = \bigcup_{t \in T_D} syn(t)$ and $Ent(t) = \bigcup_{t \in T_D} ent(t)$.

By combining the words in $Des(T_D), Syn(T_D), Ent(T_D)$, a vector, $Sem(t) = \langle f(w_1), f(w_2), \ldots, f(w_i) \rangle, w_i \in Des(T_D) \cup Syn(T_D) \cup Ent(T_D)$, is created for each term t in T_D, where the element value $f(w_i)$ for w_i is 1 if $w_i \in des(t) \cup syn(t) \cup ent(t)$, otherwise 0. With the term representation, we can group the terms based on the similarity of the term vectors. Then a clustering algorithm is applied to the vector representation to derive a set of clusters. In our experiments reported in Sect. 4, the K-medoid clustering algorithm and the Jaccard distance were used to generate the cliques. Finally, algorithm returns a set of semantic cliques $G = \{g_1, \ldots, g_n\}$, each clique consists of semantically related terms, $g_i \subseteq T_D$ as the output. The terms t in each clique $g_i \in G$ are considered as semantically related because they are grouped together based on their synonyms, definitions and entailments.

3.2   Generate Semantically Meaningful Patterns

In this section, we propose a method to interpret cliques by using the concepts in ontology, Probase [13]. The idea here is to identify groups of terms in each clique and each group can be interpreted by a set of concepts in Probase. Each group is called a semantic-based pattern.

Let C be a set of concepts in Probase, for each concept $c \in C$, Probase can retrieve a set of concepts which are related to c. For each related concept v, a conditional probability $Pr(v|c)$ is provided in Probase to measure the relatedness between v and c. For each term $t \in T_D$, let $Concept(t)$ be a set of related concepts in the ontology. Concept is a mapping defined as below:

$$Concept(T_D) \rightarrow 2^c, \quad \forall t \in T_D, \quad Concept(t) = \{c | c \in C, Pr(c|t) \neq 0\} \quad (1)$$

The top k related concepts are considered for each term $t \in T_D$. The top k concepts are the related concepts and denoted as $Concept^k(t)$, which have the highest probability values. $Concept^k(t)$ can be defined as follows;

$$Concept^k(t) = arg \, Max_{c \in Concept(t)}^k \{Pr(c|t)\} \quad (2)$$

Definition (Semantic-based Pattern and related concepts): For a clique, a semantic-based pattern of g is a set of terms, $p \subseteq g$, for any pair of terms in $p, t_i, t_j \in p$, they share some common related concepts, i.e., $Concept^k(t_i) \cap Concept^k(t_j) \neq \phi$. Semantic-based pattern of g can be defined as below:

$$p = \{t_i, t_j | t_i, t_j \in g, i \neq j, \exists c \in C, c \in Concept^k(t_i) \cap Concept^k(t_j)\} \qquad (3)$$

A set of common related concepts is called as related concepts of the pattern, as defined below.

$$Related\_concept(p) = \{c | t_i, t_j \in p, i \neq j, \exists c \in C,$$

$$c \in Concept^k(t_i) \cap Concept^k(t_j)\} \qquad (4)$$

Patterns are generated based on common matching concepts. The top k matching concepts of each term is considered. One pattern can relate to several concepts and a single semantic clique g can contain several patterns. Algorithm 2 depicts the process of generating semantic-based patterns based on related concepts in a semantic clique.

Algorithm 2 Generate patterns based on related concepts

Input   : Cliques of related terms $G = \{g_1, ..., g_n\}$
Output : A set of pattern sets $P = \{P_1, ..., P_n\}$, a set of concept sets $C_D = \{C_1, ..., C_n\}$,
         where P_i is a set of patterns and C_i is a set of concepts for clique g_i.

1. $P \leftarrow \{\}, \quad C_D \leftarrow \{\}$
2.   For each clique $g_i \in G$
3. $C_i \leftarrow \{\}$
4.   For each term $t \in g_i$
5.   Get top-k related concepts
6. $Concept^k(t) \leftarrow arg\,Max^k_{c \in Concept(t)} \{Pr(c|t)\}$
7. $C_i \leftarrow C_i \cup Concept^k(t)$
8. $P_i \leftarrow \{\}$
9. $C_c \leftarrow \{\}$ // overlapping concepts
10.   For each pair of terms $t_i, t_j \in g_i$
11.   If $Concept^k(t_i) \cap Concept^k(t_j) \neq \phi$
12. $C_c \leftarrow C_c \cup (Concept^k(t_i) \cap Concept^k(t_j))$
13.   For each $p \in P_i$
14.   If $t_i \in p$ or $t_j \in p$
15. $p \leftarrow p \cup \{t_i, t_j\}$
16.   If P_i doesn't have a pattern that contains t_i and t_j, create a new pattern
17. $q \leftarrow \{t_i, t_j\}$
18. $P_i \leftarrow P_i \cup \{q\}$
19. Return P and C_D

Semantic cliques will be the input for Algorithm 2 and the semantic based pattern generation is applied for each semantic clique $g_i \in G$. Lines 5–6 in Algorithm 2 retrieve the top k concepts for each term t in the clique g_i from the Probase. Then overlapping concepts will be considered for each pair of terms. Finally, a set of pattern sets $P = \{P_1, P_2, \ldots, P_n\}$ and the related set of concept sets $C_D = \{C_1, C_2, \ldots, C_n\}$ are generated and returned. For each clique g_i, P_i is a set of patterns and C_i is a set of related concepts for the clique. The significance here in the proposed pattern generation approach is that, it takes the semantics of the terms by considering the meanings related to the terms. Each clique can be considered as the representation of a certain topic in document collection and its corresponding patterns and concepts describe the meaning of the topic.

The following probabilities are derived to measure the pattern distribution within each clique and clique distribution (i.e., topic distribution) in the collection. Let $Pr(c|t)$ be the probability of a concept c given a term t which can be obtained from Probase, For a clique g_i, a pattern $p \in P_i$ and a concept $c \in C_i$, the probability of the concept c given the pattern p can be estimated as below:

$$Pr(c|p) = Avg_{t \in p} Pr(c|t) \tag{5}$$

The probability of the pattern p in the clique can be estimated as:

$$Pr(p|g_i) = \frac{\sum_{c \in C_i} Pr(c|p)}{\sum_{a \in g_i} \sum_{c \in C_i} Pr(c|a)} \tag{6}$$

As discussed before, each clique consists of a set of similar or related terms. In this paper, the average term frequency is used to measure the importance of the term in the collection:

$$F(t) = \frac{\sum_{d \in D} f(t,d)}{\sum_{d \in D} |d|} \tag{7}$$

$f(t,d)$ is the term frequency in d

Frequent words in a document most likely represent content of the document. Therefore, in this paper the average of term frequency in a clique is used to measure the importance of the clique:

$$F(g_i) = Avg_{t \in g_i} F(t) \tag{8}$$

Finally, the clique distribution in the collection is estimated by the normalized clique importance in the collection:

$$Pr(g_i) = \frac{F(g_i)}{\sum_{g \in G} F(g)} \tag{9}$$

4 Evaluation and Results

Semantic clique and its pattern-based representation are novel and significant features of the proposed TRuSP to generate topic representation for a given document collection. In our topic representation, we consider a semantic clique as a latent topic which represents one aspect of the collection. The topic representation generated by TRuSP can be considered as a topic model. In this section, we evaluate the effectiveness of the proposed topic representation from two aspects. Firstly, in Sect. 4.1 the quality of the topic representation will be evaluated with set of state-of-the-art topic models based on perplexity and coherence. In Sect. 4.2, we evaluate the topic representation by applying it to information filtering task. The semantic based patterns in the topic representation will be used as features for filtering relevant documents.

4.1 Evaluating the Topic Quality

Quality of the topic representation was evaluated by measuring the perplexity and the coherence of topic representation. Four existing topic modelling methods, LDA [2], MMPBTM [8], Probase-LDA [9] and CLDA [10] were chosen as the baseline models. Since MMPBTM, Probase-LDA and CLDA are also based on the LDA, experiment environment was setup with the same parameter values as $\alpha = 1$, $\beta = 0.1$ and number of iterations = 2000. For TRuSP, top $k = 5$ related concepts were chosen from Probase for each term in cliques. Reuters Corpus Volume I (RCV1) (50 collections), R8 (8 collections) and 20News groups (20 collections) datasets were used in the experiment and average values were calculated. The experiment was conducted to measure the perplexity and coherence of the topic model. Moreover, topic-word distribution was observed in different instances to evaluate the quality of topics generated by each topic model and the proposed topic representation approach.

Perplexity
Perplexity is calculated by the log likelihood of a held-out test document set. The experiment was conducted for different number of topics in LDA, MMPBTM, Probase-LDA, CLDA and TRuSP. Figure 1 depicts the perplexity score of the five approaches with variations of number of topics (or cliques for TRuSP).

The perplexity was calculated according to Eq. (10), M is the number of documents in the test set of documents in a dataset.

$$Perplexity(D_{test}) = \exp\left\{ -\frac{\sum_{d=1}^{M} \log p(W_d)}{\sum_{d=1}^{M} N_d} \right\} \tag{10}$$

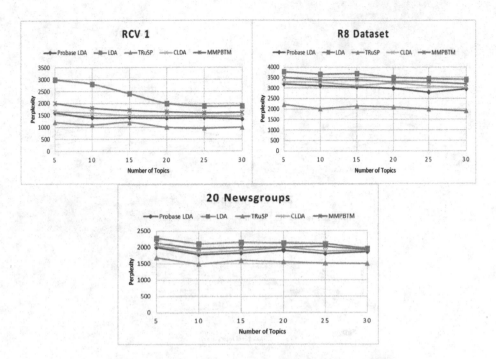

Fig. 1. Perplexity score with variations of number of topics

For LDA, MMPBTM, Probase-LDA and CLDA, $p(W_d)$ is the probability of document d, N_d is the size of d. For TRuSP, cliques are treated as topics, each topic is represented by a set of patterns, $p(W_d)$ is calculated as below and N_d is the number of patterns in d. $P = \{p_1, p_2, \ldots, p_n\}$ all patterns generated by semantic cliques.

$$P(W_d) = \prod_{p_a \in P_d} \sum_{g_i \in G} \Pr(p_a | g_i) \Pr(g_i) \tag{11}$$

Coherence

Coherence evaluates the coherent topics and how close the meanings of the topic words. We measured the topic coherence by using the method in [18]. Figure 2 shows the coherence scores of all the systems for all three datasets.

$$Coherence(t) = \sum_{m=2}^{M} \sum_{l=1}^{m-1} \log \frac{D\left(v_m^{(t)}, v_l^{(t)}\right) + 1}{D\left(v_l^{(t)}\right)} \tag{12}$$

where $v_m^{(t)}$ and $v_l^{(t)}$ are the mth and l^{th} terms within topic t, $D\left(v_m^{(t)}, v_l^{(t)}\right)$ is the co-document frequency of the two terms, and, $D\left(v_l^{(t)}\right)$ is the document frequency of the term $v_l^{(t)}$.

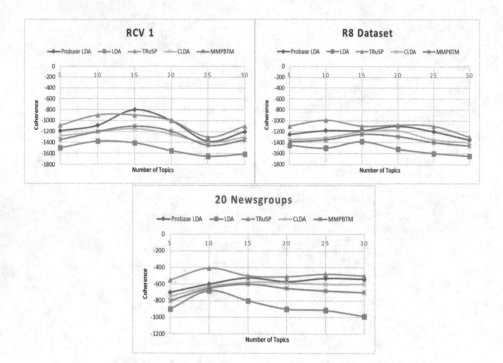

Fig. 2. Topic coherence with variations of number of topics

Measuring the topic coherence is important, since it directs towards the semantics of the topics. Topic coherence was measured for all collections of three datasets and calculated the average score with the variations of number of topics. High coherence scores indicate the high quality of the topic models. TRuSP has high coherence values in R8 and 20 Newsgroups datasets for all the occurrences of number of topics. At the same time, TRuSP has the highest coherence score for most of the number of topics variations in RCV1 dataset except for the 13–20 range. Probase-LDA has performed better than TRuSP for number of topics 13–20 range. Probase-LDA and CLDA are showing comparatively high coherence values than MMPBTM and LDA.

Observation of Topic-Word Distributions
Given that both Probase-LDA and CLDA as well as TRuSP used Probase, in this subsection, the topic-word distributions of Probase-LDA and CLDA were compared with the pattern representation of TRuSP through a few examples. In particular, we wanted to observe the topic representation of the topic number ranging from 13–20, where the topic coherence of TRuSP was slightly lower than Probase-LDA as showed in Fig. 2. In this observation, we have observed a list of topic-words in three approaches to verify the semantically meaningful topics. Sample output with first 10 topic words generated for the first 3 topics of RCV1 dataset (for collection 107, when the number of topics is = 20) is given below in Table 1.

For TRuSP, a clique is considered as a topic. Patterns generated by TRuSP for each semantic clique can be compared with the topic-words generated by Probase-LDA and CLDA. Highest ten topic words and patterns were displayed in Table 1.

Table 1. Topic words generated from Probase-LDA, CLDA and pattern representation generated from TRuSP

TRuSP			Probase-LDA		CLDA	
Semantic clique	Patterns	Annotated topic	Topic	Topic words	Topic	Topic words
G_1	{tourism, hotel}, {tourist, visitor}, {tourism, hotel, tour}, {tourism, agency}, {tourism, travel, hotel}, {holiday}, {tourism, visit, tour}, {travel}, {visit, tour, holiday}, {hotel}	Tourism	T_1	industry, million, tourist, Monday, british, oversea, australia, asian, national, bureau	T_1	ireland, trip, holiday, Monday, tourism, year, industry, irish, northern, million
G_2	{ireland, tunisia, jordan, china}, {ireland, jordan}, {irish}, {dublin}, {dublin, Europe, jordan}, {europe}, {europe, china}, {tunisia, jordan}, {ireland}, {dublin, europe}	Countries	T_2	security, real, true, relevant, concern, slow, capacity, turmoil, institution, line	T_2	tourism, visitor, year, percent, million, billion, north, visit, jordan, develop
G_3	{number, figure, output}, {number}, {number, percent}, {number, cent}, {number, data, percent}, {billion}, {data, cent, percent}, {output}, {million, billion}, {cent}	Number/representation	T_3	performance, credit, advertisement, total, resolution, number, contrast, earlier, republic, minister	T_3	korea, travel, industry, europe trip, international, minister, market israel, jakarta

The topic words in T_1 in Probase LDA and CLDA are correspondent to the pattern-words in semantic clique G_1. According to our observation all the words in the semantic patterns and semantic cliques are highly related to each other when compared to the topic words generated by Probase -LDA and CLDA. Further, the semantic cliques generated by the TRuSP can be easily categorized into topics as given in the annotated topics in the Table 1.

4.2   Information Filtering Based Evaluation

In this section, the generated semantically meaningful patterns were used as features to represent the documents in the training collection (i.e., representing the information needs of the user). The features are used to calculate a score for determining the relevance of a new document for information filtering. The accuracy and the relevancy of the patterns for information filtering were evaluated by comparing with seven baseline systems in the information filtering domain. Reuters Corpus Volume I (RCV1)

[17] dataset was used. Seven baseline models were used in this evaluation which cover term-based, phrase-based, and pattern-based features to represent user information needs for information filtering.

For a new document d, its relevance to the collection is measured by the following score based on the clique distribution and the patterns in the document:

$$\text{Score(d)} = \sum_{g \in G} \left(\sum_{P \in g,d} \Pr(P|g) \right) * \Pr(g) \tag{13}$$

The top 20, precision, recall, F-measure and Mean Average Precision (MAP) values were calculated for collections in the RCV 1 dataset. Top 20 documents were derived based on the $Score(d)$ of documents and $F\ measure$ is calculated with the precision and recall where, $F\ measure = \frac{2*precision*recall}{precision + recall}$. Mean Average Precision (MAP) value is calculated by getting the mean of average precision value for each collection. The average results of the 50 collections of RCV1 are shown below in the Table 2. It is clearly noticed that the results of the semantic based pattern generation approach (TRuSP) are promising and better than all the other approaches.

Table 2. Comparison of Coefficients from Atomistic (RCV1 dataset)

	Top 20	F-measure	MAP
LDA-words	0.458	0.426	0.421
TNG [15]	0.446	0.386	0.374
N-Gram	0.401	0.386	0.361
Frequent Closed Patterns -FCP	0.428	0.385	0.362
Probase-LDA [9]	0.258	0.252	0.286
MMPBTM [8]	0.552	0.460	0.478
CLDA [10]	0.239	0.317	0.259
TRuSP	0.561	0.472	0.481

5 Discussion

According to the experimental results, TRuSP contains the lowest perplexity score for all three datasets. Hence it proves the quality of topic representation in TRuSP is better than the other approaches. Probase-LDA has the next lowest score which is lower than the LDA, MMPBTM and CLDA. TRuSP had the highest coherence scores for both R8 and 20Newsgroups datasets. For RCV1 dataset, TRuSP has high coherence values in most of the occurrences other than the range of number of topics 13–20. This might happen due to increasing the size of semantic cliques. Then less related words will be grouped within the semantic cliques when the size gets increased. Even though the coherence values of TRuSP is lower than the Probase-LDA within the topics range 13–20, it is observed that the generated topics of TRuSP is more meaningful than all the other approaches in all the occurrences. Probase-LDA and CLDA also have comparatively high coherence scores respectively even though most of the time less than

TRuSP. TRuSP, CLDA and Probase-LDA have used the concepts in ontology Probase to determine the topics and have high coherence values compared to LDA and MMPBTM. It shows the importance of using semantic concepts in topic modelling. Probase-LDA used Probase for conceptualization, that aimed to calculate the Dirichlet priors for LDA and concepts were not directly involved in the output. In CLDA also concepts are identified using the Probase but concepts are considered as a layer on top of the traditional LDA. Meanings and relationships among the terms were not properly considered in both approaches. Nevertheless, the topic coherence of LDA is the lowest and it might happen due to not considering the conceptual meanings of the content. Generating semantic cliques based on WordNet and grouping semantic based patterns based on Probase to represent the training document collection can improve the performance of information filtering. Thus, the content goes through two semantic based phases to enrich the outcome and therefore the information filtering-based evaluation depicts a considerably high performance. MMPBTM has the next highest performance since it has combined the pattern mining and topic modeling techniques together whereas the remaining baseline models consider the term-based (LDA, Probase-LDA, CLDA), pattern-based (FCP) and phrases-based (TNG) topic modeling approaches.

6 Conclusions

Extracting the meaningless words and generating irrelevant topics are two major problems in topic modelling. Semantics of the text data are important to grasp the real meaning of the content. TRuSP is a combination of rich features of concepts and relevancy of the terms, which focuses to derive meaningful set of topics based on semantic patterns. Generating semantic cliques from the related terms and generating patterns based on the matching concepts are the main components in this approach. Experimental results concluded that the results of TRuSP were stronger than the baseline models. Generated semantic patterns can be used as the main construct in topic modeling and topics can be generated based on the matching concepts. The novelty here in this approach is that, it identifies semantically related terms based on general knowledge bases (i.e., WordNet and Probase) to represent the topics in a document collection. Handling the words which are not matching with WordNet and Probase can be defined as a main challenge of TRuSP and in future, we plan to handle the non-matching words using our own approach.

References

1. Evans, J.A., Aceves, P.: Machine translation: mining text for social theory. Ann. Rev. Sociol. **42**(1), 21–50 (2016)
2. Blei, D.M., Ng, A.Y., Jordan, M.I.: Latent Dirichlet allocation. J. Mach. Learn. Res. **3**, 993–1022 (2003)
3. Blei, D.M.: Probabilistic topic models. Commun. ACM **427**, 77–84 (2012)

4. Hofmann, T.: Probabilistic latent semantic indexing. In: Proceedings of the 22nd Annual International ACM SIGIR Conference on Research and Development in Information Retrieval - SIGIR 1999. ACM Press (1999)
5. Le, Q.V., Mikolov, T.: Distributed Representations of Sentences and Documents ([n. d.])
6. Mikolov, T., Chen, K., Corrado, G., Dean, J.: Efficient estimation of word representations in vector space ([n. d.])
7. Andrzejewski, D., Zhu, X., Craven, M.: Incorporating domain knowledge into topic modeling via Dirichlet forest priors. In: ICML, pp. 25–32. ACM (2009)
8. Gao, Y., Xu, Y., Li, Y.: Pattern-based topics for document modelling in information filtering. IEEE Trans. Knowl. Data Eng. **27**(6), 1629–1642 (2015)
9. Yao, L., Zhang, Y., Wei, B., Qian, H., Wang, Y.: Incorporating probabilistic knowledge into topic models. In: Cao, T., Lim, E.-P., Zhou, Z.-H., Ho, T.-B., Cheung, D., Motoda, H. (eds.) PAKDD 2015. LNCS (LNAI), vol. 9078, pp. 586–597. Springer, Cham (2015). https://doi.org/10.1007/978-3-319-18032-8_46
10. Tang, Y.K., Mao, X.L., Huang, H., Shi, X., Wen, G.: Conceptualization topic modeling. Multimed. Tools Appl. **77**, 3455–3471 (2017)
11. Gao, Y., Li, Y., Lau, R.Y.K., Xu, Y., Bashar, M.A.: Finding semantically valid and relevant topics by association-based topic selection model. ACM Trans. Intelligent Syst. Technol. **9**(1), 1–22 (2017)
12. Miller, G.A.: WordNet: a lexical database for English. Commun. ACM **38**, 39–41 (1995)
13. Wu, W., Li, H., Wang, H., Zhu, K.Q.: Probase. In: Proceedings of International Conference on Management of Data - SIGMOD 2012. ACM Press (2012)
14. Landauer, T.K., Foltz, P.W., Laham, D.: An introduction to latent semantic analysis. Discourse Process. **25**, 259–284 (1998)
15. Wang, X., McCallum, A., Wei, X.: Topical N-grams: phrase and topic discovery, with an application to information retrieval. In: Seventh IEEE International Conference on Data Mining (2007)
16. Chen, Z., Mukherjee, A., Liu, B., Hsu, M., Castellanos, M., Ghosh, R.: Discovering coherent topics using general knowledge. In: CIKM, pp. 209–218. ACM (2013)
17. Lewis, D.D., Yang, Y., Rose, T.G., Li, F.: RCV1: a new benchmark collection for text categorization research. J. Mach. Learn. Res. **5**, 361–397 (2004)
18. Mimno, D., Wallach, H.M., Talley, E., Leenders, M., McCallum, A.: Optimizing semantic coherence in topic models. In: Proceedings of the Conference on Empirical Methods in Natural Language Processing, pp. 262–272 (2011)

Outlier Detection Based Accurate Geocoding of Historical Addresses

Nishadi Kirielle$^{(\boxtimes)}$, Peter Christen, and Thilina Ranbaduge

Research School of Computer Science, The Australian National University,
Canberra, ACT 2600, Australia
nishadi.kirielle@anu.edu.au

Abstract. Research in the social sciences is increasingly based on large and complex databases, such as historical birth, marriage, death, and census records. Such databases can be analyzed individually to investigate, for example, changes in education, health, and emigration over time. Many of these historical databases contain addresses, and assigning geographical locations (latitude and longitude), the process known as *geocoding*, will provide the foundation to facilitate a wide range of studies based on spatial data analysis. Furthermore, geocoded records can be employed to enhance record linkage processes, where family trees for whole populations can be constructed. However, a challenging aspect when geocoding historical addresses is that these might have changed over time and therefore are only partially or not at all available in modern geocoding systems. In this paper, we present a novel method to geocode historical addresses where we use an online geocoding service to initially retrieve geocodes for historical addresses. For those addresses where multiple geocodes are returned, we employ outlier detection to improve the accuracy of locations assigned to addresses, while for addresses where no geocode was found, for example due to spelling variations, we employ approximate string matching to identify the most likely correct spelling along with the corresponding geocode. Experiments on two real historical data sets, one from Scotland and the other from Finland, show that our method can reduce the number of addresses with multiple geocodes by over 80% and increase the number of addresses from no to a single geocode by up to 31% compared to an online geocoding service.

Keywords: Geocode matching · String comparison · Open Street Map

1 Introduction

The recent surge in the digitization of historical records, such as censuses, and birth, death, and marriage certificates, is enabling social and health scientists to explore human behavioral patterns across time at an unprecedented level of detail [8]. The economic, social, medical, and demographic history of people has been the interest that led to the growth of historical data analysis [7,18]. In this context, geospatial analysis plays an important role to uncover a great

© Springer Nature Singapore Pte Ltd. 2019
T. D. Le et al. (Eds.): AusDM 2019, CCIS 1127, pp. 41–53, 2019.
https://doi.org/10.1007/978-981-15-1699-3_4

Table 1. Sample historical addresses and the corresponding retrieved modern addresses using Open Street Maps along with their geocodes and address types.

Historical address	Modern address	Geocode [latitude, longitude]	Address type
Kilmorie	No matching addresses	No geocode	–
Kilmore	Kilmore, Highland Scotland, IV44 8RG, UK	[57.0942387, −5.8720672]	Hamlet
Feorlig	Feorlig, Highland Scotland, IV55 8ZL, UK	[57.4020757, −6.4979426]	Hamlet
	Feorlig, A863, Feorlig Highland, IV55 8ZL, UK	[57.4052835, −6.4974833]	Post box

deal of hidden patterns in populations using geographical information such as residential addresses that are commonly available in historical databases.

The process of geocoding aims to assign a geographic location (latitude and longitude) to a textual address string [2]. In order to obtain accurate geocodes for a large number of addresses, a comprehensive reference database consisting of addresses and their locations is required. Alternatively, online services, such as Google Maps or Open Street Maps (OSM) [10], some of which provide an application programming interface (API), can be employed for geocoding.

While geocoding modern addresses generally results in accurate locations being assigned to address strings [2,16], the process of geocoding historical addresses is quite challenging. This is due to address quality issues and differences between historical addresses and the addresses available in modern geocode reference databases or geocoding services. Address quality issues can occur because of spelling variations, missing values, and incomplete addresses [6], and many historical addresses do not follow modern address structures. For instance, evidence in the Digitising Scotland project [6] suggests that Nineteenth century addresses in census records mostly only provide township names [15]. This is in contrast to commonly used modern hierarchical address structures that generally consist of street numbers and names, postcodes, and town names [4].

Due to such imperfections in historical addresses, querying such address strings using modern geocoding services commonly leads to partial matches with multiple contemporary addresses regardless of the high quality, coverage, and efficiency of the used geocoding service or geocoding reference database. As a result, when querying historical addresses, existing geocoding services will return either a single, multiple, or an empty set of locations.

Table 1 shows an example of historical addresses from a real-world database we use in our experiments in Sect. 4. These addresses are extracted from Nineteenth century birth certificates from the Isle of Skye in Scotland [15]. Also shown are the results when geocoding these historical address strings using OSM, which returns no address, or a list of one or several matching contemporary addresses, their geocodes, and their corresponding address types.

Prior research in geocoding historical addresses involves establishing a separate *gazetteer* (geographical dictionary) by associating historical addresses with geocodes using existing gazetteer sources [4,13,20]. This approach, however, does not facilitate geocoding historical addresses in the absence of corresponding gazetteers with associated geocodes. To the best of our knowledge, no previous studies have investigated how to incorporate modern geocoding services such as OSM to geocode historical addresses.

Contribution. We examine how to best utilize modern geocoding services to geocode historical addresses, and propose a novel geocoding method that uses Open Street Maps (OSM) [10] to geocode historical addresses. We employ two refinement phases to find a single geocode for those addresses where OSM returns either multiple or no geocodes in the first phase: We use outlier detection to find the most likely location for addresses that have multiple geocodes returned from OSM; and apply approximate string matching for address strings that did not receive any geocode to identify the most similar corresponding address string along with its geocode. We evaluate our method on two historical data sets showing how it can lead to significantly improved geocoding results compared to applying a basic online geocode service such as OSM only.

2   Related Work

We now describe research related to our work, including approaches to geocoding of historical addresses as well as the use of geocoding for record linkage.

St-Hilaire et al. [20] presented a historical address geocoding approach for Canadian census manuscripts from 1911 to 1951 by implementing a reference gazetteer which associates historical addresses with geocodes. Rather than geocoding at the level of addresses, the authors have geocoded at the level of census subdivisions (CSD), a small unit for which census returns were published, by associating each address with the corresponding CSD polygon as per the historical records. Due to variations in addresses over time, the CSD polygons are generated separately for each year by referencing and overlaying 2001 Statistics Canada digital maps onto historical maps. Logan et al. [13] have geocoded US census records from 1880 with a resolution of street-level addresses by associating street level historical addresses with contemporary TIGER (Topologically Integrated Geographic Encoding and Referencing) files which comprise geospatial information released by the US Census Bureau.

A recent approach by Lafreniere et al. [11] has implemented a framework for geocoding historical addresses using an address point locator created for each historical period by combining historical sources. All historical sources with images have been georeferenced using ArcGIS, a modern geocoding service. A similar study by Cura et al. [4] relaxes the need of complete gazetteers and instead employs geohistorical objects which contain information extracted from historical sources for the process of geocoding.

In 2015, Daras et al. [5] proposed a framework for geocoding historical addresses in the Digitising Scotland project [6]. In contrast to the previous work

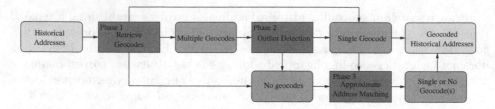

Fig. 1. Overview of geocoding historical addresses consisting of (1) geocode retrieval, (2) processing of addresses with multiple geocodes with outlier detection, and (3) approximate string matching for addresses with no geocode. The blue boxes indicate phases while the red boxes indicate intermediate results. Yellow boxes indicate input and output. (Color figure online)

that used historical gazetteers, the authors employed exact and fuzzy string matching to map historical addresses to modern addresses. This framework compares historical to modern addresses and employs manual clerical review to geocode addresses that do not map to a corresponding modern address.

In the context of record linkage, several attempts have been proposed to incorporate geographical information when linking databases. Blakely et al. [1] have utilized geocodes in the blocking step when linking New Zealand census to mortality data. Schraagen and Kosters [19] employed distance-based measures as consistency constraints applied on graphs for family reconstruction. More recently, in a genealogical network inferring algorithm proposed by Malmi et al. [14], the authors used a probabilistic record linkage model to construct family trees with attribute similarity features including a geographical distance.

Overall, these studies have shown that existing geocoding approaches for historical addresses are highly data dependent, where most of the research work uses existing gazetteers to conduct geocode matching. What is not yet clear is the impact of using available online geocoding services, such as OSM, for the geocoding process of historical addresses.

3   Geocoding Historical Addresses

In this section, we present our method to geocode historical addresses, as outlined in Fig. 1 and summarized next. The aim of geocoding historical addresses is to find a single and accurate geographical location for each address.

The first phase, as described in Sect. 3.1, involves retrieving geocodes from an online geocoding service. In our work, we utilize the freely accessible geocoding service OSM [10]. We denote with $\mathbf{A}$ the set of unique historical addresses for which we are interested in finding geocodes. The retrieval of geocodes from the online geocoding service can result in three subsets: (1) addresses with a single geocode, $\mathbf{A}_S$, (2) addresses with multiple geocodes, $\mathbf{A}_M$, and (3) addresses with no geocodes, $\mathbf{A}_N$, where $\mathbf{A} = \mathbf{A}_S \cup \mathbf{A}_M \cup \mathbf{A}_N$. Addresses in $\mathbf{A}_M$ and $\mathbf{A}_N$ require further processing to obtain a valid single geocode of their locations.

Over time, conventions in address structures have evolved, and as a consequence, historical addresses often do not follow the hierarchical structure of contemporary addresses [4]. Existing historical sources might also only contain incomplete or partial addresses due to choices and inefficiencies in the digitizing processes. Accordingly, for a particular historical address, multiple matching contemporary addresses (each with a different location) may be returned. In the second phase of our method, as we describe in Sect. 3.2, we process the addresses in $\mathbf{A}_M$. We use the type of each matching contemporary address, such as *building*, *village*, *hamlet*, or *camping area*, and the geographical distances between the geocodes for a given address to remove likely irrelevant geocodes.

As a result of spelling variations and differences between historical and contemporary addresses, existing geocoding services potentially do not contain location information for all historical addresses [4]. The third phase of our method, described in Sect. 3.3, therefore focuses on identifying the most similar correct spelling variation in $\mathbf{A}_S$ for the addresses in $\mathbf{A}_N$, assuming that those addresses contain spelling variations or missing tokens from their correct version, and assigning corresponding geocodes to the addresses in $\mathbf{A}_N$.

3.1 Retrieving Geocodes

The retrieval of geocodes for a historical address from a geocoding service is straight-forward when the queried address string returns either a single or multiple geocode(s). However, due to data quality issues in historical addresses discussed above, there are instances where a historical address string cannot be matched to any existing address known to the geocoding service. For historical addresses that comprise of multiple words (tokens), in the absence of any geocode for the full address, we tokenize the address (split a string at whitespaces) and obtain geocodes for different subsets of tokens to generalize the address.

However, the hierarchical inconsistency and incompleteness of historical addresses can complicate the process of identifying hierarchical information in an address. Therefore, we iteratively remove each token (starting from the first) and query the geocoding service with the remaining set of tokens. For instance, the address 'Brae Stein Waternish' can be queried with 'Stein Waternish', 'Brae Waternish' and 'Brae Stein', and we then consider the union of geocodes retrieved from all three queries. If multiple matching contemporary addresses with multiple geocodes are returned for these queries, further processing (as described next) can help to obtain the best matching geocode.

3.2 Geocoding Historical Addresses with Multiple Geocodes

The second phase of our method, as detailed in Algorithm 1, focuses on processing each address in $\mathbf{A}_M$ to obtain a single valid geocode by removing one or more invalid geocodes. We use geographical distances between geocodes, types of contemporary addresses associated with each geocode, and an outlier detection based function to filter out invalid geocodes.

Depending on the application, we can decide which type of addresses, $\mathbf{T}$, to consider when multiple modern addresses are returned for a particular historical address string. For example, we consider an order of addresses of type $\mathbf{T} = [village, hamlet, residential\ area, building]$ for the experimental evaluation as we are interested in places where people live. In the presence of addresses of types $\mathbf{T}$, we filter the addresses with the highest priority type in lines 3 to 5.

Algorithm 1: *Processing multiple geocodes (Phase 2)*

Input:
- $\mathbf{A}_M$:   Set of historical addresses with multiple geocodes
- $\mathbf{A}_S$:   Set of historical addresses with a single geocode
- $\mathbf{A}_N$:   Set of historical addresses with no geocode
- $\mathbf{T}$:   List of appropriate address types ordered according to their priority
- t_{min}:   Threshold for minimum distance between a valid set of geocodes
- t_{max}:   Threshold for maximum distance between a set of geocodes with no outliers
- f: Outlier detection function
- z: Threshold for the outlier detection function

Output:
- $\mathbf{A}_S$:   Set of historical addresses with a single geocode
- $\mathbf{A}_N$:   Set of historical addresses with no geocode

```
1:  for a ∈ A_M do:                                        // Loop over addresses
2:      g = a.geocode_set                    // Get initial geocode set for address a
3:      g_T = GetPriorityGeocodes(g, T)   // Get set of geocodes filtered by priority types
4:      if g_T ≠ ∅ then:               // Check if prioritized geocodes are available
5:          g = g_T      // Update geocode set with the prioritized geocodes if available
6:      d_min = GetMinimumDistance(g)   // Retrieve minimum distance among geocodes
7:      if d_min > t_min then:                          // Check minimum distance
8:          A_N = A_N ∪ {a}          // All geocodes are too far apart, add address to set A_N
9:      else:
10:         d_max = GetMaximumDistance(g)     // Get maximum distance among geocodes
11:         if (d_max > t_max) and (|g| > 2) then:       // Check possibility of outliers
12:             g = GetOutlierRemovedGeocodes(g, f, z)   // Get outlier removed geocodes set
13:         if g ≠ ∅ then:           // Check if geocode set is not empty after outlier removal
14:             a.geocode = GetAverageGeocode(g)            // Get average geocode
15:             A_S = A_S ∪ {a}          // Add to set of addresses with single geocode
16:         else:
17:             A_N = A_N ∪ {a}          // Add to set of addresses with no geocode
18: return A_S, A_N
```

For a given address, we validate the set of geocodes by exploiting the minimum and maximum geographical distance between them. If the minimum distance, d_{min}, between any geocode pair in its set $\mathbf{g}$ is above a certain threshold t_{min}, then these geocodes are geographically too scattered. As we are employing an unsupervised process, it is not possible to decide which geocode is correct in a scattered set of geocodes. We therefore consider them as an invalid set of geocodes and add the address to $\mathbf{A}_N$ (lines 6 to 8).

In lines 9 to 17, we then aim to identify any outlying geocodes in the set of geocodes for a given address. We only employ outlier detection if the maximum distance, d_{max}, between any pair of geocodes in $\mathbf{g}$ is greater than the threshold t_{max} (lines 10 to 12). The thresholds t_{min} and t_{max} can be set by the user based on the expected circular proximity of a valid set of geocodes.

We employ outlier detection to identify any geocodes that are far away from others for a given address. Because the number of multiple geocodes for a given address is usually small, and the set of geocodes are not a set of numerical values, we use modified versions of standard statistical outlier detection functions such as the z-score normalization [9] and the robust variation of z-score normalization [17]. In z-score normalization, if a data point deviates more than z standard deviations from the mean of the data distribution, it is considered as an outlier. The values for z used vary in the range of $2 \leq z \leq 4$ [12]. The robust variation of z-score replaces the mean and standard deviation in the normalization process with the median and median absolute deviation to avoid the effect of outliers on the statistical measures [17].

To apply these statistical outlier detection functions in the context of geocodes, we use the distances between all geocodes in the set $\mathbf{g}$ for a given address, rather than the geocodes themselves. The pair-wise distances between geocodes are calculated using the great circle distance, which reflects the shortest distance between two points on the Earth measured along the surface using the Haversine equation [21]. Let us assume the radius of Earth is R and the geocodes g_1 and g_2 have longitude and latitude values as (x_1, y_1) and (x_2, y_2), respectively. Then the distance $d_{1,2}$ between these two geocodes can be calculated as:

$$h_{1,2} = \sin^2\left(\frac{x_2 - x_1}{2}\right) + \cos(x_1) \times \cos(x_2) \times \sin^2\left(\frac{y_2 - y_1}{2}\right)$$
$$d_{1,2} = R \times 2 \times \arcsin(\min(1, \sqrt{h_{1,2}})) \tag{1}$$

Now let us define the set $\mathbf{D}$ as the $n(n-1)/2$ pair-wise distances calculated between the n geocodes in $\mathbf{g}$, with $n = |\mathbf{g}|$, returned for one given address, where $n > 2$. For a given geocode $g_i \in \mathbf{g}$, we denote its set of distances to all other geocodes in $\mathbf{g}$ as $\mathbf{D}_i$. We calculate the average of a set of distances as avg(), the standard deviation as std(), the median as med(), and the median absolute deviation as mad(). The z-score, z_i, and robust z-score, rz_i, for geocode g_i are then calculated as:

$$z_i = \frac{|avg(\mathbf{D}) - avg(\mathbf{D}_i)|}{std(\mathbf{D})} \qquad rz_i = \frac{|med(\mathbf{D}) - med(\mathbf{D}_i)|}{mad(\mathbf{D})} \tag{2}$$

If the value z_i or rz_i is greater than the predefined threshold value, z, then geocode g_i is considered as an outlier.

After outliers are identified and removed (lines 11 and 12 in Algorithm 1), the remaining geocodes are averaged to obtain a single geocode for the given historical address (we use the average instead of the median due to the generally very small numbers of geocodes in $\mathbf{g}$). The computational complexity of Algorithm 1 is $O(|\mathbf{A}_M| \cdot g^2)$, where g is the average size of the sets of geocodes, $\mathbf{g}$.

3.3   Geocoding Historical Addresses with No Geocodes

In the third phase of our method, as outlined by Algorithm 2, we employ approximate string matching to identify the most similar address string for the addresses in $\mathbf{A}_N$ where no geocode was found. To identify the most similar address string, we can either use addresses for which we have already identified a single geocode, $\mathbf{A}_S$, or alternatively use existing historical gazetteers [4, 13, 20].

Algorithm 2 requires a string similarity function $sim()$, a similarity threshold value s_t, and two sets of addresses: those without geocodes, $\mathbf{A}_N$ (possibly due to spelling variations), and those for which a single geocode is available, $\mathbf{A}_S$. For each address $a_N \in \mathbf{A}_N$, if the highest similarity score s_{max} of a_N with any address $a_S \in \mathbf{A}_S$ is above the similarity threshold s_t (which decides if two addresses are matching or not), then we assign the geocode of the best matching address, a_S^{best}, to the non-geocoded address a_N in line 9. Otherwise, the geocode of a_N is left as unknown in $\mathbf{A}_N$ and kept for manual review. The computational complexity of Algorithm 2 is $O(|\mathbf{A}_N| \cdot |\mathbf{A}_S|)$.

Algorithm 2: *Approximate Address Matching (Phase 3)*

Input:
- $\mathbf{A}_N$:   Set of historical addresses with no geocodes
- $\mathbf{A}_S$:   Set of historical addresses with a single geocode
- $sim()$:   String similarity function
- s_t: Threshold for string similarity calculation

Output:
- $\mathbf{A}_N$:   Set of historical addresses with no geocodes
- $\mathbf{A}_S$:   Set of historical addresses with a single geocode

```
 1: for a_N ∈ A_N do:                          // Loop over non-geocoded addresses
 2:    s_max = 0                           // Initialize maximum similarity value to 0
 3:    for a_S ∈ A_S do:                          // Loop over geocoded addresses
 4:       s = sim(a_N, a_S)              // Calculate the similarity between addresses
 5:       if s ≥ s_max then:       // Check if similarity is above the maximum similarity
 6:          s_max = s                        // Update the highest similarity score
 7:          a_S^best = a_S                       // Update the most similar record
 8:    if s_max ≥ s_t then: // Check if the maximum similarity score is above the threshold
 9:       a_N.geocode = a_S^best.geocode      // Assign the geocode of most similar address
10:    A_N = A_N \ {a_N}          // Remove the record from non-geocoded address set
11:    A_S = A_S ∪ {a_N}                 // Add the record to geocoded address set
12: return A_S, A_N
```

Many different approximate string similarity functions have been developed [3]. One commonly used such function specific for English names is Jaro-Winkler [22]. This function calculates a similarity between 0 (strings are completely different) and 1 (strings are the same) by counting the numbers of common and transposed characters. Given address strings commonly contain several tokens, we adapted this comparison function where we first sort all tokens in the addresses to be compared and then apply Jaro-Winkler on the sorted tokens. A set of pre-experiments showed good results using this approach. However, our method can use any string similarity function to match addresses. Deciding on the string similarity threshold depends on the expected similarity of matching addresses.

4   Experimental Evaluation

We evaluated our method to geocoding historical addresses on two real data sets. The Scottish (Isle of Skye) data set[1] [15] consists of 17,614 birth records from the

---

[1] Not publicly available, for similar data see: https://www.scottish-places.info.

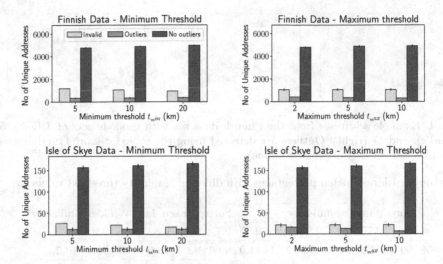

Fig. 2. Variation of multiple geocoded address categories with respect to different t_{min} and t_{max} values (as discussed in Sect. 3.2).

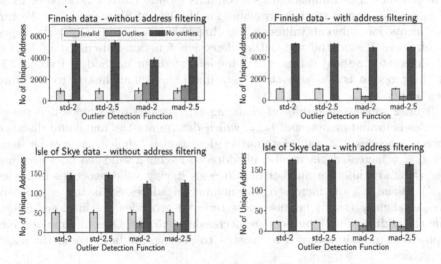

Fig. 3. Variation of multiple geocoded address categories with respect to different outlier detection functions f and thresholds z (as discussed in Sect. 3.2).

Isle of Skye from 1861 to 1901 with 1,268 unique addresses. The Finnish data set[2] [14] contains 4,962,236 birth records from 1600 to 1917, with only 9,392 unique addresses (most of these only the name of a hamlet or village). For this data set we therefore considered the combination of village and parish names as the full address because the same village name commonly occurs across dif-

[2] Available at: http://hiski.genealogia.fi/hiski?en.

Fig. 4. Example addresses from the Finnish data set with geocode sets of 3 (left), 5 (middle), and 17 (right). Outliers are detected using the robust z-score function and shown as red stars. (Color figure online)

Table 2. Address match percentages with different similarity threshold values s_t.

	Jaro-Winkler similarity						Sorted token Jaro-Winkler similarity					
s_t	0.7	0.75	0.8	0.85	0.9	0.95	0.7	0.75	0.8	0.85	0.9	0.95
%	99.4	94.2	74.1	49.5	29.4	11.0	99.0	93.5	73.6	49.7	29.4	10.5

ferent parishes. The Finnish data set contains ground truth locations for most addresses. No ground truth is available for the Isle of Skye data set. We ran experiments for different values of the thresholds, t_{min} and t_{max}, the types of addresses, **T**, and the two outlier detection functions discussed in Eq. (2) with different threshold values, z. We implemented our method in Python 2.7, and the program is available at: https://dmm.anu.edu.au/histrl/ to facilitate repeatability.

Figure 2 shows the results of changing t_{min}, which determines if a set of geocodes is invalid or not, and t_{max}, which determines the maximum distance between geocodes in a set before outlier detection is applied. As can be seen, when t_{min} is increased, the number of addresses having invalid geocodes becomes lower ($\approx$32%) while the number of addresses having valid geocodes increases ($\approx$4%). When t_{max} is increased, the number of addresses having no outliers increases slightly ($\approx$4%) because most received geocodes are in closer range, while the addresses with outliers are decreasing ($\approx$57%). Overall, however, our proposed method is robust with regard to settings of both these threshold parameters.

Figure 3 shows results of using the two different outlier detection functions to identify outlying geocodes. The maximum number of multiple geocodes retrieved for an address was 40 for both data sets, while the average and median were 4.2 and 2, respectively. Because of these small numbers of geocodes for each address, the robust z-score function, using median, tends to perform better in identifying outliers compared to the average based z-score function. Furthermore, the figure provides strong evidence of the significance of using address type filtering. The numbers of invalid addresses and addresses with outliers are considerably higher if no address type filtering is applied compared to with address type filtering. This is because the geocodes are selected for each address for the highest priority address type while other geocodes that are unlikely to match are removed.

Table 3. Summary of geocoded addresses after geocoding with OSM, after applying our geocoding algorithm and an analysis of proximity with ground truth.

	Finnish	Isle of Skye
Total number of unique address strings	9,392	1,268
Number of addresses with a single geocode from OSM	2,654	298
Number of addresses with multiple geocodes from OSM	6,268	195
Number of addresses with no geocodes from OSM	470	775
Number of addresses with multiple geocodes resulted in:		
A valid geocode with the proposed method	$5,283 \pm 207$	159 ± 15
An invalid geocode with the proposed method	985 ± 207	36 ± 15
Number of addresses with no geocodes resulted in:		
A valid geocode with the proposed method	9 ± 6	331 ± 191
No geocode with the proposed method	$1,447 \pm 204$	479 ± 191
Ground truth analysis:		
Number of addresses with a valid geocode within 1 km	$2,284 \pm 175$	–
Number of addresses with a valid geocode within 5 km	$3,226 \pm 127$	–
Number of addresses with a valid geocode within 10 km	$3,682 \pm 138$	–
Number of addresses with a valid geocode within 20 km	$4,065 \pm 126$	–

Figure 4 shows examples of how robust z-score correctly identifies outliers of geocode sets of different sizes.

We evaluated the effect of approximate string matching and the similarity threshold s_t, as discussed in Sect. 3.3, using the Isle of Skye data set. Table 2 shows a clear decrease in the number of matches when s_t is increased. However, a higher s_t more likely identifies the correct variation of a misspelled address due to the high similarity between the pair of address strings.

Finally, Table 3 presents the number of unique addresses of the two data sets when geocoded only with OSM, and the averages and standard deviations when different parameter settings of our geocoding method are applied. As can be seen, our method is able to find a valid single geocode for over 80% of addresses when multiple geocodes were retrieved from OSM. The approximate string matching phase is also capable of identifying correct spelling variations for misspelled addresses and identify a single valid geocode for no geocoded addresses for up to 31% for the Isle of Skye data set. However, as the Finnish data set is normalized to contain unique addresses, in the third phase our method is unable to identify valid spelling variations for most of the Finnish addresses.

The final section of Table 3 shows the number of addresses located with a calculated geocode within 1, 5, 10, and 20 km when compared with the ground truth location. Due to the variations in geocoding datums and the accuracy of OSM, the proximity of calculated geocodes and ground truth geocodes varies.

5   Conclusion

We have presented a novel method to geocoding historical addresses using an online geocoding service. We apply outlier detection and approximate string matching to identify accurate locations for those addresses where multiple or no geocodes were retrieved. Our evaluation on two real historical data sets showed significant improvements in geocoding historical addresses using our method compared to an online geocoding service. As future work, we plan to improve the geocode retrieval from an online geocoding service by recognizing the tokens in addresses using Hidden Markov model-based approaches [3], and we aim to compare our method with prior methods. We also aim to explore how a suitable threshold for outlier detection can be learned from the data and explore alternative outlier detection functions for geocoding of historical addresses.

Acknowledgements. This work was partially funded by the ARC under DP160101934. We like to thank Alice Reid, Ros Davies and Eilidh Garrett for their work on the Isle of Skye data set, especially Eilidh for her helpful advice on historical demography of the Isle of Skye.

References

1. Blakely, T., Woodward, A., Salmond, C.: Anonymous linkage of New Zealand mortality and census data. ANZ J. Public Health **24**(1), 92–95 (2000)
2. Christen, P., Churches, T., Willmore, A.: A probabilistic geocoding system based on a national address file. In: Australasian Data Mining Conference, Cairns (2004)
3. Christen, P.: Data Matching. Springer, Heidelberg (2012). https://doi.org/10.1007/978-3-642-31164-2
4. Cura, R., Dumenieu, B., Abadie, N., et al.: Historical collaborative geocoding. ISPRS Int. J. Geo-Inf. **7**(7), 262 (2018)
5. Daras, K., Feng, Z., Dibben, C.: HAG-GIS: a spatial framework for geocoding historical addresses. In: GIS Research UK Conference, Leeds (2015)
6. Dibben, C., Williamson, L., Huang, Z.: Digitising Scotland (2012). http://gtr.rcuk.ac.uk/projects?ref=ES/K00574X/2
7. Garrett, E., Reid, A.: Introducing 'movers' into community reconstructions: linking civil registers of vital events to local and national census data: a Scottish experiment. In: Bloothooft, G., Christen, P., Mandemakers, K., Schraagen, M. (eds.) Population Reconstruction, pp. 263–283. Springer, Cham (2015). https://doi.org/10.1007/978-3-319-19884-2_13
8. Georgala, K., van der Burgh, B., Meeng, M., Knobbe, A.: Record linkage in medieval and early modern text. In: Bloothooft, G., Christen, P., Mandemakers, K., Schraagen, M. (eds.) Population Reconstruction, pp. 173–195. Springer, Cham (2015). https://doi.org/10.1007/978-3-319-19884-2_9
9. Grubbs, F.: Procedures for detecting outlying observations in samples. Technometrics **11**(1), 1–21 (1969)
10. Haklay, M., Weber, P.: OpenStreetMap: user-generated street maps. IEEE Pervasive Comput. **7**(4), 12–18 (2008)
11. Lafreniere, D., Gilliland, J.: All the World's a Stage: a GIS framework for recreating personal time-space from qualitative and quantitative sources. Trans. GIS **19**(2), 225–246 (2015)

12. Ley, C., Klein, O., et al.: Detecting outliers: do not use standard deviation around the mean, use absolute deviation around the median. J. Exp. Soc. Psychol. **49**(4), 764–766 (2013)
13. Logan, J., Jindrich, J., Shin, H., Zhang, W.: Mapping America in 1880: the urban transition historical GIS project. Hist. Methods **44**(1), 49–60 (2011)
14. Malmi, E., Gionis, A., Solin, A.: Computationally inferred genealogical networks uncover long-term trends in assortative mating. In: World Wide Web Conference, Lyon (2018)
15. Reid, A., Davies, R., Garrett, E.: Nineteenth-century Scottish demographyfrom linked censuses and civil registers. Hist. Comput. **14**(1–2), 61–86 (2002)
16. Roongpiboonsopit, D., Karimi, H.: Quality assessment of online street and rooftop geocoding services. Cartogr. Geogr. Inf. Sci. **37**(4), 301–318 (2010)
17. Rousseeuw, P., Hubert, M.: Robust statistics for outlier detection. Wiley Interdisc. Rev.: Data Min. Knowl. Discov. **1**(1), 73–79 (2011)
18. Ruggles, S., Fitch, C.A., Roberts, E.: Historical census record linkage. Annu. Rev. Sociol. **44**(1), 19–37 (2018)
19. Schraagen, M., Kosters, W.: Record linkage using graph consistency. In: Perner, P. (ed.) MLDM 2014. LNCS (LNAI), vol. 8556, pp. 471–483. Springer, Cham (2014). https://doi.org/10.1007/978-3-319-08979-9_36
20. St-Hilaire, M., Moldofsky, B., Richard, L., Beaudry, M.: Geocoding and mapping historical census data: the geographical component of the Canadian Century Research Infrastructure. Hist. Methods **40**(2), 76–91 (2007)
21. Van Brummelen, G.: Heavenly Mathematics: The Forgotten Art of Spherical Trigonometry. Princeton University Press, Princeton (2012)
22. Winkler, W.: String comparator metrics and enhanced decision rules in the Fellegi-Sunter model of record linkage. In: Survey Research Methods ASA (1990)

SPDF: Set Probabilistic Distance Features for Prediction of Population Health Outcomes via Social Media

Hung Nguyen[1]([✉]), Duc Thanh Nguyen[2], and Thin Nguyen[1]

[1] Applied Artificial Intelligence Institute, Deakin University, Geelong, Australia
hung@deakin.edu.au
[2] School of Information Technology, Deakin University, Geelong, Australia

Abstract. Measurement of population health outcomes is critical to understanding the health status of communities and thus enabling the development of appropriate health-care programmes for the communities. This task acquires the prediction of population health status to be fast and accurate yet scalable to different population sizes. To satisfy these requirements, this paper proposes a method for automatic prediction of population health outcomes from social media using Set Probabilistic Distance Features (SPDF). The proposed SPDF are mid-level features built upon the similarity in posting patterns between populations. Our proposed SPDF hold several advantages. Firstly, they can be applied to various low-level features. Secondly, our SPDF fit well problems with weakly labelled data, i.e., only the labels of sets are available while the labels of sets' elements are not explicitly provided. We thoroughly evaluate our approach in the task of prediction of health indices of counties in the US via a large-scale dataset collected from Twitter. We also apply our proposed SPDF to two different textual features including latent topics and linguistic styles. We conduct two case studies: across-year vs across-county prediction. The performance of the approach is validated against the Behavioral Risk Factor Surveillance System surveys. Experimental results show that the proposed approach achieves state-of-the-art performance on linguistic style features in prediction of all health indices and in both case studies.

Keywords: Population health · Social media

1 Introduction

The societal-level health measures, such as self-perceived mental and physical health, or mortality, provide indicators representing the health situation of a community, and hence inform subsequent public health promotion planning [15, 20]. These would also help governments develop appropriate health-care programmes for the communities. Typically, the health outcomes of a community are summarised either from the health status of individuals in that community or represented as the distribution of individual health outcome measures [15].

© Springer Nature Singapore Pte Ltd. 2019
T. D. Le et al. (Eds.): AusDM 2019, CCIS 1127, pp. 54–63, 2019.
https://doi.org/10.1007/978-981-15-1699-3_5

Traditionally, input data for population health analysis of a community is collected from individuals in that community through surveys. These surveys can be done via means such as telephone interviews or postal feedback for pre-designed questionnaires. The strength of the traditional approach is twofold: reliable answers could be obtained as the questionnaires have been designed by experts, and the population of interest could be actively targeted and hence feedback would be relevant and representative for the population. However, organising such kind of surveys is usually costly, time-consuming, and not scalable. It often takes years to make surveyed results publicly available. For instance, the Behavioral Risk Factor Surveillance System (BRFSS) reports in 2017 were collected in or before 2015[1]. In addition, small portions of the population participating in surveys might not properly reflect holistic characteristics of the entire population. For instance, BRFSS – the largest health survey dataset conducted in the world – includes just about 400,000 interviews each year.

Thanks to the advent of online social networking platforms – also known as "social media", a new source of data namely social media data has opened numerous crowdsourcing-based applications, including health-related applications. Billions of people are daily connecting in social networks such as Facebook, Twitter, Instagram, Flickr, etc. to share their wellness status with private contacts or the public. Moreover, the advances of the Internet and mobile technologies allow people, easily and seamlessly, post their activities and events in real-time manners. Social media data therefore potentially captures individual's health status, both mental and physical, via the mood, thinking, activities and communications made by members in populations [4]. With spatial and temporal information, social media data provides an effective way to automatically determine populations and track their changes over time. In addition, vast amount of data can be collected through social media more rapidly and easily yet at much lower cost than traditional surveys [7].

Health-related data can be captured from a wide range of sources including medication, demographics, genetics, social activity, family history, life style, environment and social relationships, and would help expand knowledge in disease diagnosis, prevention, treatment and management [1]. Recent studies in health analysis via social media have examined various ways to extract the health-related content in social media and exhibited the potential of the approach. These studies indicate a strong correlation between what people in a community post and the health outcomes of the community [13]. Linguistic Inquiry and Word Count (LIWC) [16], and a latent topics model such as latent Dirichlet allocation (LDA) [2], are commonly used to extract the health-related content from textual data, e.g., tweets. However, existing population health analysis methods, e.g., [3,13,18], treat populations independently and individually while there may exist similarities in the way members of different populations communicate in social media and these similarities may be important indicators of the health status of the populations. For instance, similar populations may indicate similar health outcomes.

---

[1] https://www.cdc.gov/brfss/.

Towards this direction, we investigate the similarities between populations for prediction of the health outcomes of the populations. In particular, we propose novel features namely Set Probabilistic Distance Features (SPDF) which are built upon set distance, i.e., distance between sets. In our approach, populations are considered as sets and tweets in populations are considered as elements of sets. The advantages of our approach are twofold. Firstly, the proposed SPDF are mid-level features and can be applied to various low-level features. Secondly, our approach does not require the labels (i.e., health status) for individual tweets. Instead, only the health status of populations are considered. This make the approach adaptive to weakly labelled data, i.e., only the labels of sets are available while the labels of sets' elements are not explicitly provided. We apply the proposed SPDF on two different textual features including LIWC and latent topics. We also experiment our approach with various distance metrics. The proposed approach is evaluated in the task of prediction of health indices of counties in the US via a large-scale dataset collected from Twitter. Two case studies: across-year and across-county prediction of population health outcomes are conducted. Experimental results show the potential of the proposed approach in both case studies and the approach achieves state-of-the-art performance when SPDF is applied to linguistic style in prediction of all health outcomes and in both case studies.

The remainder of our paper is organized as follows. In Sect. 2 we briefly review related work on social media-based population health study. In Sect. 3 we present our approach. We describe our experiments and analyse results in Sect. 4. Finally we provide some remarks and conclude the paper in Sect. 5.

2   Related Work

Social media has found a broad range of applications such as ecology and environmental management, transportation or epidemiology [6,19]. Most people today are connecting to social networks using their mobile phones to share the places they visit, what they do and how they feel on the go. This communication mode provides huge volumes of spatio-temporal, user generated data which also open novel opportunities for health-related applications. Recently, social media has significantly enhanced healthcare research due to the offered advantages including novel and effective lens to keep track of user activities and their interactions in a real-time manner. Prior studies in public health analysis through social media have utilised tweet data obtained from social media at population scale and demonstrated their potentials. An early work is the Google Flu Trends [9], developed for detecting influenza epidemics for more than 25 countries. This work is motivated from the observation that the frequency of patients' queries on physician visits is highly correlated to the level of weekly influenza-like symptoms. Accumulated geotagged information from social media data can also be harnessed to determine spatial health-related issues, monitor the spread of infectious diseases, and/or analyse the effect of a concept on public health. For instance, geospatial social media data can be used to specify geographic densities of clinical concepts in regions of interest [8,11], spatially cluster groups of data having similar characteristics [17], or build recommendation systems to advice locations of interest [22]. Similarly, the relationship between urban form and well-being can be identified from social media [21]. In addition, location information from social media is valuable for mapping clinical contents to spatial representations [10].

To deal with the problem of social media based data analysis at population levels, it is crucial to characterise the population-level content. Several techniques have been proposed for textual-based population representation. A straight-forward approach is to aggregate textual features extracted from individuals of a population to form the population-level features for that population. For instance, in [3,18], tweets in a county is concatenated into a so-called aggregated tweet on which features are extracted. This approach is simple to implement and aims to alleviate computational burden in processing large-scale data. However, the approach does not consider the relationships between features at individual-level which may potentially convey predictive information of population health outcomes.

In [14], so-called kernel-based features are constructed on top of textual features using kernel functions. Kernel functions capture statistical correlations between textual features. However, this method does not take into account the relationship between populations (e.g., the commonality in the way people in different populations make their posts). Since only the heath status of populations is given while the health status of each individual post in populations is not provided, the similarity between populations may indicate the similarity between health outcomes. Another drawback of the method in [14] is that the dimension of kernel-based features is quadratic of the dimension of texture features which are used in the kernel functions. This makes the method suffer from high computational complexity.

3   Proposed Method

As argued in our introductory section, the similarity between populations may indicate the similarity between health outcomes. This observation inspires us to develop population-level features that capture the relationships between populations for data analysis at population scale. Specially, we propose here Set Probabilistic Distance Features (SPDF) which are formed from the distance between sets. In our paper, populations are considered as sets in which tweets in populations are considered as elements of sets. Our approach offers two advantages. Firstly, the proposed SPDF are mid-level features and can be adapted to various low-level features. Secondly, our approach fits well data analysis tasks with weakly labelled data. For instance, the labels (health status) for individual tweets are not given but only the health status of populations are.

The crucial part of our proposed SPDF is the distance metric used to measure the similarity between populations (sets). There exist several set distance metrics and Hausdorff distance is probably the most popular one. However, Hausdorff distance favours extreme cases. In particular, the Hausdorff distance between two sets is biased by the maximal distance between elements of these two sets. To overcome this shortcoming, our proposed SPDF defines the distance between two sets in a probabilistic way that takes into account the distances between all pairs of elements in two sets. The proposed SPDF thus is able to capture the commonality of the sets.

To begin with, let us introduce notations that are used in our formulation. Let C denote a county from which a set of tweets t_i is collected. For simplicity, we denote $C = \{t_i\}$. A county is considered as a "set" and its tweets are considered as "elements" in the set. A tweet t is encoded by a feature vector $f(t) \in \mathbb{R}^N$. We define the likelihood $P(t|C)$ that a tweet t is drawn from a county (i.e. set of tweets) C by using the following marginalisation,

$$
\begin{aligned}
P(t|C) &= \sum_{t_i \in C} P(t|t_i) \\
&= \sum_{t_i \in C} exp\left(\frac{-d^2(f(t), f(t_j))}{\sigma^2}\right)
\end{aligned}
\tag{1}
$$

where $d(f(t), f(t_j))$ is a certain distance between the feature vectors $f(t)$ and $f(t_j)$. In our experiments, we implemented $P(t|C)$ with various distance measures such as Euclidean, city block, cosine or correlation.

We then define a new probabilistic similarity between two counties C_j and C_k as

$$
S(C_i, C_j) = \sum_{t \in C_i} P(t|C_j) + \sum_{t \in C_j} P(t|C_i)
\tag{2}
$$

where $P(t|C_j)$ and $P(t|C_i)$ is defined in (1). As shown, the greater $S(C_i, C_j)$ is, the more similar C_i to C_j is and vice versa. We note that our proposed probabilistic similarity somehow capture the commonality of sets.

Given a training set including counties whose the health indices are annotated (labelled), we first apply the K-means algorithm on the training set to obtain clusters. Note that our clustering is different from the traditional K-means method. In particular, each data point in our case is a county and different counties have different number of tweets. In addition, the centroids of clusters are not the means of clusters. Instead, clusters are initialised by randomly partitioning the training set. The centroid of each cluster is a dummy county which is formed by aggregating the tweets of all the counties in that cluster. Comparing a county with a cluster centroid is performed using (2). The clustering step results in a set of K centroids denoted as $\{C_1, C_2, ..., C_K\}$.

For each training/test county C, we encode this county by a feature vector,

$$
\mathcal{V}(C) = \langle S(C, C_1), S(C, C_2), ..., S(C, C_K)\rangle
\tag{3}
$$

where $S(C, C_i)$ is defined in (2).

Finally, the SPDF vectors generated on the training set are used to train a regression model, for instance, the Kernel Ridge Regression [12]. For prediction of health outcomes, each test county is encoded by a SPDF vector that is fed to the regression model for estimating the county health index.

4 Experiments

4.1 Data

We crawled 1,129,928,183 tweets made in years 2014, 2015, and 2016, and geotagged with US latitude/longitude coordinates, which were mapped to US geocodes. We also collected 152,853,038 tweets made in 2013 for learning latent topics. The LDA algorithm proposed in [2] was adopted to learn latent topics. To reduce computational complexity, we limited the number of tweets for each county in each year to 100,000 tweets selected randomly. Consequently, 448,567,987 tweets were collected.

The collected geotagged tweets were associated with the US counties by mapping their geocodes to the Federal Information Processing System (FIPS) codes using the cartographic boundary files provided by the US Census Bureau in 2013. There were 3,221 different geocodes corresponding to 3,221 counties in the US. Note that we used only tweets with associated latitude/longitude coordinates, those with self-reported location information but without coordinates were not considered in our study.

We used the BRFSS survey reports as the ground truth for our experiments. The surveys were conducted by the Centers for Disease Control and Prevention (CDC) via telephone and collected data from the US residents regarding to heir health-related risk behaviours, chronic health conditions, and health outcomes. BRFSS contains about 400,000 interviews completed each year[2] and is currently the largest health survey system, not only in the US but also in the world. The questionnaires in BRFSS surveys are categorised into core sections including current health status, number of healthy days, inadequate sleep, chronic health conditions, and optional modules such as healthcare access or social context.

For estimating health indices case study, we used the annual health ranking data of counties in BRFSS surveys including (i) poor or fair health – percent of adults that report fair or poor health, (ii) poor physical health days – the average number of reported physically unhealthy days per month, and (iii) poor mental health days – the average number of reported mentally unhealthy days per month. For classifying health situation case study, bottom and top 500 healthiest counties for each health index[3] were used.

4.2 Experimental Setup

Case Studies. To evaluate our proposed approach, we run experiments on two case studies: across-year and across-county prediction. For across-year prediction, the data of year 2014 is used for training a regression model, then we

---

[2] https://www.cdc.gov/brfss/index.html.
[3] https://www.usnews.com/news/healthiest-communities/rankings.

validate the approach on the data of years 2015 and 2016. For across-county prediction, the dataset of the same year is split in to training set (70%) and test set (30%). We measure the performance of prediction of health outcomes using Spearman's rank correlation coefficient between predicted values and the ground truth values (from the BRFSS reports).

We validate our proposed SPDF on two common textual features including latent topics [2] and linguistic style features (LIWC) [16]. The numbers of latent topics and linguistic style categories are both set to 78, resulting in a 78-dimensional feature vector for each tweet. Recall that the SPDF is built based on distance metrics, i.e., $d(.,.)$ in Eq. (1). In our experiments, we evaluate our approach with different distance metrics including cityblock, correlation, Euclidean, and cosine. For regression model, Kernel Ridge Regression algorithm – linear least squares with l2-norm regularization – is adopted. To get the best model, grid search cross-validation with 5-fold cross validation is applied.

We also compare our approach with the baseline method in [3,18].

Computational Resources. Various preprocessing tasks in our experiments such as data aggregation, linguistic feature extraction (from billions of tweets), and matrix distance are at large-scale thus require high computational cost. We employed Spark on top of Hadoop [23] for these tasks. Spark is a cluster computing platform which enables distributed and parallel computations on a cluster scaled up to 8,000 nodes. Additionally, Spark is an in-memory based system that keeps data in memory for convenient subsequent iterations, thus allows much faster computations than disk-based systems like Hadoop MapReduce [5]. Specifically, the entire Spark Hadoop cluster comprises 8 CentOS 7.2 physical machines, each equipped with Intel® Xeon® E5-26700 (8 cores, 16 threads) CPU, 128 GB RAM, Intel Xeon Phi Coprocessor (60 cores), and 24TB HDD.

4.3   Results and Discussion

Across-Year Prediction. Table 1 shows prediction results of the across-year scenario on topic features. As shown, in comparison to the average distance, the proposed approach achieves significant improvement. More specifically, in prediction of generic health index of 2015, our approach yields a correlation at 0.66 with the ground truth when cityblock distance is used. The largest improvement of 22% is also obtained in predicting generic health index of 2016 with cityblock distance. Further, the improvement is obtained on predicting all health indices in both years 2015 and 2016. We observe that on average, the correlation distance results in the best performance (Spearman's rho = 0.48), while cosine distance shows the poorest one (Spearman's rho = 0.38).

The prediction results of the same scenario but on LIWC features are shown in Table 2. In this case, our approach with Euclidean distance consistently show the best performance in predicting all indices. In comparison to the baseline, 16% of improvement is obtained in estimating the generic and physical indices of 2016. However, correlation and cosine distances appear to be ineffective metrics on LIWC features.

Table 1. Performance (in Spearman's rho) on predicting the US county health indices for the across-year prediction scenario. Topic features are extracted from tweets. Best results are in **bold**.

	2015			2016		
	Generic	Mental	Physical	Generic	Mental	Physical
Baseline	0.60	0.54	0.48	0.33	0.32	0.30
Cityblock	**0.66**	0.48	0.50	**0.55**	0.23	0.24
Correlation	0.61	0.52	0.59	0.37	0.32	**0.48**
Euclidean	0.57	**0.56**	0.56	0.32	**0.36**	0.36
Cosine	0.61	0.51	**0.60**	0.24	0.18	0.19

Table 2. Performance (in Spearman's rho) on predicting the US county health indices for the across-year prediction scenario. LIWC features are extracted from tweets. Best results are in **bold**.

	2015			2016		
	Generic	Mental	Physical	Generic	Mental	Physical
Baseline	0.53	0.42	0.45	0.29	0.22	0.22
Correlation	0.47	0.39	0.41	0.25	0.21	0.19
Cosine	0.49	0.37	0.39	0.27	0.18	0.18
Euclidean	**0.56**	**0.46**	**0.49**	**0.45**	**0.31**	**0.38**

Table 3. Performance (in Spearman's rho) on predicting the US county health indices for the across-county prediction scenario with topic features. Best results are in **bold**.

	2015			2016		
	Generic	Mental	Physical	Generic	Mental	Physical
Baseline	0.66	0.62	**0.62**	0.60	**0.61**	0.57
Cityblock	0.65	0.61	0.59	0.61	0.55	0.55
Correlation	**0.67**	**0.63**	**0.62**	**0.62**	0.60	**0.58**
Euclidean	0.62	0.58	0.59	0.57	0.50	0.50
Cosine	0.66	**0.63**	0.61	0.60	0.59	0.57

Across-County Prediction. Table 3 shows prediction results on the across-county case study on various distances. In this case we only investigate the use of latent topics. As shown in our experiments, among all the distances, correlation performs best in most cases while Euclidean performs worst in estimating all indices. On average, the proposed approach with correlation distance gains 2.7% performance improvement compared to the baseline.

To conclude, the proposed approach outperforms the baseline in both prediction scenarios and on both types of low-level features LIWC and topics. Further, from experimental results we observe that the correlation distance is superior to other metrics on topics, while the Euclidean is dominant on LIWC features.

5 Conclusion

Population heath outcome measurement is critical for governments to develop health promotion strategies and learn what initiatives and programs work best. In this paper we propose Set Probabilistic Distance Features (SPDF) to encode population health information. We evaluated our proposed features on the task of predicting the health indices of the US counties in two case studies: across-year and across-county prediction and measured its performance against the BRFSS survey reports. Experimental results show that our approach not only obtains significant correlation with the ground truth, but also out performs the baseline method. As social media platforms have become the most popular means for people to connect and share what is happening with them, our results suggest that it is feasible to estimate general health outcomes at societal level through the lens of social media in a real-time and low-cost manner. Applying SPDF to different types of social media data will be our future work.

References

1. Andreu-Perez, J., Poon, C.C.Y., Merrifield, R.D., Wong, S.T.C., Yang, G.-Z.: Big data for health. IEEE J. Biomed. Health Inform. **19**(4), 1193–1208 (2015)
2. Blei, D.M., Ng, A.Y., Jordan, M.I.: Latent Dirichlet allocation. J. Mach. Learn. Res. **3**, 993–1022 (2003)
3. Culotta, A.: Estimating county health statistics with Twitter. In: Proceedings of the SIGCHI Conference on Human Factors in Computing Systems, pp. 1335–1344 (2014)
4. De Choudhury, M., Gamon, M., Counts, S., Horvitz, E.: Predicting depression via social media. In: Proceedings of the International AAAI Conference on Weblogs and Social Media (ICWSM), pp. 128–137 (2013)
5. Dittrich, J., Quiané-Ruiz, J.-A.: Efficient big data processing in Hadoop MapReduce. Proc. VLDB Endow. **5**(12), 2014–2015 (2012)
6. Dredze, M.: How social media will change public health. IEEE Intell. Syst. **27**(4), 81–84 (2012)
7. Dredze, M., Paul, M.J.: Natural language processing for health and social media. IEEE Intell. Syst. **29**(2), 64–67 (2014)
8. França, U., Sayama, H., McSwiggen, C., Daneshvar, R., Bar-Yam, Y.: Visualizing the "Heartbeat" of a city with Tweets. Complexity **21**(6), 280–287 (2016)
9. Ginsberg, J., Mohebbi, M.H., Patel, R.S., Brammer, L., Smolinski, M.S., Brilliant, L.: Detecting influenza epidemics using search engine query data. Nature **457**(7232), 1012–1014 (2009)
10. Lan, R., Lieberman, M.D., Samet, H.: The picture of health: Map-based, collaborative spatio-temporal disease tracking. In: Proceedings of the SIGSPATIAL International Workshop on Use of GIS in Public Health, pp. 27–35 (2012)
11. Leetaru, K., Wang, S., Cao, G., Padmanabhan, A., Shook, E.: Mapping the global Twitter heartbeat: the geography of Twitter. First Monday **18**(5) (2013)

12. Murphy, K.P.: Machine Learning: A Probabilistic Perspective. MIT Press, Cambridge (2012)
13. Nguyen, T., et al.: Kernel-based features for predicting population health indices from geocoded social media data. Decis. Support Syst. **102**, 22–31 (2017)
14. Nguyen, T., et al.: Prediction of population health indices from social media using kernel-based textual and temporal features. In: Proceedings of the International Conference on World Wide Web Companion, pp. 99–107 (2017)
15. Parrish, R.G.: Peer reviewed: measuring population health outcomes. Prev. Chronic Dis. **7**(4) (2010)
16. Pennebaker, J.W., Booth, R.J., Boyd, R.L., Francis, M.E.: Linguistic Inquiry and Word Count: LIWC 2015 [Computer software]. Pennebaker Conglomerates Inc. (2015)
17. Quercia, D., Capra, L., Crowcroft, J.: The social world of Twitter: topics, geography, and emotions. In: Proceedings of the International AAAI Conference on Weblogs and Social Media (ICWSM), vol. 12, pp. 298–305 (2012)
18. Schwartz, H.A., et al.: Characterizing geographic variation in well-being using tweets. In: Proceedings of the International AAAI Conference on Web and Social Media (ICWSM), pp. 583–591 (2013)
19. Shekhar, S., et al.: Spatiotemporal data mining: a computational perspective. ISPRS Int. J. Geo-Inf. **4**(4), 2306–2338 (2015)
20. Thacker, S.B., Stroup, D.F., Carande-Kulis, V., Marks, J.S., Roy, K., Gerberding, J.L.: Measuring the public's health. Public Health Rep. **121**(1), 14–22 (2006)
21. Venerandi, A., Quattrone, G., Capra, L.: City form and well-being: what makes London neighborhoods good places to live? In: Proceedings of the SIGSPATIAL International Conference on Advances in Geographic Information Systems (2016)
22. Ye, M., Yin, P., Lee, W.-C.: Location recommendation for location-based social networks. In: Proceedings of the SIGSPATIAL International Conference on Advances in Geographic Information Systems, pp. 458–461 (2010)
23. Zaharia, M., et al.: Fast and interactive analytics over Hadoop data with Spark. Usenix Login **37**(4), 45–51 (2012)

Estimating County Health Indices
Using Graph Neural Networks

Hung Nguyen[1]([✉]), Duc Thanh Nguyen[2], and Thin Nguyen[1]

[1] Applied Artificial Intelligence Institute, Deakin University, Geelong, Australia
hung@deakin.edu.au
[2] School of Information Technology, Deakin University, Geelong, Australia

Abstract. Population health analytics is fundamental to developing responsive public health promotion programs. A traditional method to interpret health statistics at population level is analyzing data aggregated from individuals, typically through telephone surveys. Recent studies have found that social media can be utilized as an alternative population health surveillance system, providing quality and timely data at virtually no cost. In this paper, we further investigate the use of social media to the task of population health estimation, based on a graph neural network approach. Specifically, we first introduce a graph modeling method to construct the representation of each county as a graph of interactions between health-related features in the community. We then adopt a graph neural network model to learn the population health representation, ended by a regression layer, to estimate the health indices. We validate our proposed method by large-scale experiments on Twitter data for the task of predicting health indices of the US counties. Empirical results show a significant correlation with the reported health statistics, up to a Spearman correlation coefficient (ρ) value of 0.69, and that our graph-based approach outperforms the existing methods. These promising results also suggest potential application of graph-based models to a range of societal-level analytics tasks through social media.

Keywords: Social networks · Social web and applications · Population health · Graph neural networks

1 Introduction

Population health statistics is fundamental to public health policy formulation. It helps understand societal-level health behaviors, identify underlying health concerns and develop responsive public healthcare programs. For this purpose, many governments have conducted studies to track public health trends, such as epidemiology or overall health status in populations. Conventional sources of health data include medical reports, physical examinations, personal interviews and phone/mail surveys. In the US, for example, the Behavioral Risk Factor Surveillance System (BRFSS) is a premier system of telephone surveys that

© Springer Nature Singapore Pte Ltd. 2019
T. D. Le et al. (Eds.): AusDM 2019, CCIS 1127, pp. 64–76, 2019.
https://doi.org/10.1007/978-981-15-1699-3_6

collect state data about health risk behaviors, chronic health conditions, and use of preventive services among US residents.[1] The results from the BRFSS have been used by the US Centers for Disease Control and Prevention (CDC) to plan public health programs at both local and national levels.

Organizing a conventional health survey, such as telephone interviews, is expensive and time-consuming, despite the high-quality data collected. It often takes years to make results publicly available; for instance, the BRFSS 2017 results are typically based on data collected in or before 2015.[2] Furthermore, due to the limited number of participants, these traditional methods might not reflect properly public health outcomes. For example, BRFSS, as the largest health survey ever conducted in the world, completes about 450,000 interviews each year. These limitations give rise to the need for an alternative solution to collect comprehensive and timely population health statistics.

Social media has recently become an important "sensor" to track health behaviors and trends in real-time. Billions of people are daily using social network platforms such as Facebook, Twitter, or Flickr to share their status, photos or opinions with public or their connected friends. Additionally, social media provides innovative data such as geolocation, user responses or networks, which previously unavailable. As such, to some extent, social media reflects individual's health status, both mental and physical, via mood, thinking, activities and communication [9]. In comparison with traditional surveys, social media provides larger, novel and diverse amounts of data [12]. Furthermore, data can be collected from social media rapidly at low cost.

Various studies have shown the potential application of social media to the task of tracking health behaviors at population scale. Significant results have been found in tracking chronic illnesses [7], detecting depression or mental well-being [1,3,8,9,15,26], or estimating generic and physical health outcomes [18]. A common approach in these studies is to analyze user linguistic data, in particular Twitter messages, based on the link between language and health [6,13,25]. To this end, the linguistic features, such as LIWC lexicon [23], are extracted from aggregated data, then a machine learning algorithm is adopted to learn the prediction model. For instance, in [7,28], health-related representations for a population were obtained by simply aggregating individual tweets. In [17,18], high-level features capturing statistical relationships between low-level features (i.e., LIWC and latent topics) were proposed.

In this paper, we investigate the application of graph neural networks to the task of population health estimation through social media. Specifically, from the aggregated data, we first extract low-level linguistic features, then construct a representation of each population as a graph of interactions between the low-level features. Consequently, each county is represented as a graph where the nodes are the low-level features and the edges encode the relationships between them. Finally, we adopt a state-of-the-art graph neural network model to learn health index prediction models.

---

[1] https://www.cdc.gov/chronicdisease/resources/publications/aag/brfss.htm.
[2] https://www.cdc.gov/brfss/

The rest of the paper is organized as follows. In Sect. 2 we briefly review related work on social media-based health study and representations for population health analytics. In Sect. 3 we propose graph-based features for population health from interactions between social media data and describe new model for estimating population health outcomes. We then describe dataset and experimental setup, results and discussion in Sect. 4. Finally we state some potential future work and conclude the paper in Sect. 5.

2   Related Work

2.1   Social Media for Population Health Analytics

Apart from the feasibility of monitoring individual health-related concerns, social media could also contributed to better understanding population health perspective, forming a new field known as digital epidemiology. Online generated content, when harnessed appropriately, can provide information about disease and health dynamics in populations around the world [27]. For example, a probabilistic model was developed to train on a large corpus of Twitter posts, which have been shared by individuals diagnosed with clinical depression, to determine if the posts could indicate depression [8]. The model took advantage of social activity, emotion and language manifested as signals reflecting individual mental health status. Using this model, De Choudhury et al. [8] introduced a "social media depression index" which can characterize levels of depression in populations. Paul and Dredze [20] applied the Ailment Topic Aspect Model (ATAM) [21] to model behavior around a variety of diseases of importance in public health on over 1.5 million health-related tweets and discovered mentions of dozens ailments. The results showed qualitative correlations with public health data and evaluations of model output. In addition, prior knowledge was incorporated into this model to several tasks, including localizing illnesses by geographic region and measuring behavioral risk factors [20].

Social media has also exhibited promising performance in public health estimation. Nguyen et al. proposed kernel-based features for predicting the US counties health indices from geocoded tweets and gained significantly higher prediction performance than did the existing techniques by up to 16.3% [18]. The mid-level kernel-based features, generated by considering the distributions of textual features at population level along with the relationships of textual features between individuals, can also be applied to applications requiring data analytics at population levels [18].

2.2   Features for Population Health Analytics

Primary Textual Features. Most social media-based population health analysis approaches, to date, extract semantic information from textual content such as status, tags or comments. Textual features, such as linguistic style or latent topics, are therefore often used as the discriminators.

Linguistic Style. A meaningful connection between language and health out-comes has been discovered in [22]. For example, when reading depressing stories, judges tend to get depressed accordingly. Based on these findings, a software package, named Linguistic Inquiry and Word Count (LIWC), was developed to extract psycho-linguistic features for a given text [24]. The LIWC goes through every word of document, makes comparison the word with a dictionary built by [23], then calculates the percentage of each LIWC category, finally lists all categories and the rates that each category was used in the given text.

Various studies have adopted LIWC features to problems of social media based health analysis. Culotta performed a linguistic analysis of the Twitter activity to estimate health-related statistics from the County Health Rankings & Roadmaps projects, including health outcomes [7]. The experimental results show significant correlation with 6 of the 27 health statistics, in which the LIWC lexicon outperformed alternatives [7]. Tweets has been indicated providing better representation of a community's health than demographic variables alone [7]. In addition, the linguistic features were found to be predictive of the subjective well-being of the US counties [28].

Features for Health Study at Population Level. For social media based data analysis at population levels, it is crucial to characterize a population from its individual data points. Several techniques have been proposed for textual-based population representation which can be categorized into two approaches: aggregation and statistics.

Aggregation for Population-Scale Features. To characterize a population, a straight-forward approach is aggregation, in which a population feature set is formed from data concatenated from all of its collected data points. This simpli-fied method alleviates computational cost of big data processing, however might ignore significant characteristics of a population. Particularly, it lacks of the dis-tribution of features over the population and the relationships between features which might convey predictive information of population health outcomes and consequently decrease the prediction accuracy.

Aggregation was applied in [7,28], where the collected tweets in a county were concatenated and then input into a feature extraction procedure.

Statistical Textual Features. The kernel-based features [18] are constructed on top of low-level features (e.g., LIWC or LDA) by computing their correlation via a kernel function. In particular, a set of kernel-based features is computed from a "kernel function" to generate statistical features for a population.

2.3   Graph Neural Networks

Recent success of deep convolutional neural networks (CNN) in computer vision, speech recognition and natural language processing has led to the idea of extend-ing the convolution operation to graph structures. Existing approaches can be categorized into two main streams: spectral and spatial. Methods in spectral approach, e.g., [5,10], are based on spectral graph theory [4,30]. In this approach, convolution operations are performed in spectral domain and thus graphs are required to be homogeneous.

In contrast, the spatial approach performs convolution operations directly on original graphs [2,14,19]. This technique enables learning on heterogeneous graphs and can be applied to both node and graph classification task. Two major challenges of non-spectral approaches are (1) defining the convolution operation with differently sized neighborhoods and (2) designing a pooling technique to enable hierarchical graph representation learning.

3   Method

Inspired by recent success in graph neural networks, in this work we propose a novel, graph-based model for the task of population health prediction. Interactions in social media data may reflect behaviors of communities and thus may contribute to the discrimination of health status of populations. In this work, we first describe how to model interactions in social media data using graph theory. We then construct graph-based representation of a population from its collected social media data. Finally, we employ a deep learning algorithm on the graphs to train a population health prediction model. Our proposed method is graphically illustrated in Fig. 1.

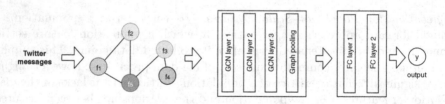

Fig. 1. Illustration of graph-based model for estimating population health through social media.

3.1   Graph Construction

We consider interactions between linguistic low-level features in social media data. In particular, we focus on the most commonly used features in the task of population health prediction, namely LIWC [23].

Formally, let $T^P = \{t_1^P, ..., t_N^P\}$ denote the set of tweets collected from a population P. Suppose that each tweet $t_i^P \in T^P$ can be described by a d-dimensional feature vector $\mathbf{f}^P{}_i = \left[f_{i,1}^P, ..., f_{i,d}^P\right] \in \mathbb{R}^d$, e.g., $d = 78$ psycho-linguistic features in LIWC. Having the low-level features $\{\mathbf{f}^P{}_i\}$, we define the interaction $I_{j,k}^P$ between two arbitrary features j and k using the radial basis function as,

$$I_{j,k}^P = \exp\left[-\left(\frac{1}{N}\sum_{i=1}^{N} f_{i,j}^P - \frac{1}{N}\sum_{i=1}^{N} f_{i,k}^P\right)^2 \Big/ 2\sigma^2\right] \tag{1}$$

The interaction $I_{j,k}^P$ defined in (1) represents the consistency in the variation of the average values of feature j and feature k in all tweets collected in the population P. The more similar the tendency of features j and k is, the higher the interaction $I_{j,k}^P$ is.

We represent the interactions between features in P via a graph $G^P\left(V^P, E^P\right)$ where $V^P = \left\{v_1^P, ..., v_d^P\right\}$ is the set of vertices, each vertex corresponds to a feature and E^P is the set of undirected edges defined as,

$$E^P = \left\{(v_j^P, v_k^P) \in V^P \times V^P \,\middle|\, I_{j,k}^P > \theta\right\} \tag{2}$$

where θ is a user-defined threshold.

3.2   Graph-Based Population Health Prediction Model

Having population representations as graphs as described above, the task of population health estimation now turns out to be a graph regression problem. Our model takes the input as a graph which has been constructed as described previously, then applies a graph neural network model to learn a representation vector. In order to evaluate the effectiveness of graph-based method, we investigate a number of recently success graph neural network models, including graph convolutional networks (GCN) [14], graph attention networks [29], graph isomorphism network (GIN) [31], and a combined GAT-GCN architecture. The detail of each graph neural network architecture are described as follows:

GCN-Based Graph Representation Learning. GCN model [14] was originally designed for the problem of semi-supervised node classification. The model enables to learn hidden layer representations that capture both local graph structures and features of nodes. This fits well our constructed undirected, node attributed graphs. To make the GCN applicable to the task of learning a representation vector of the whole graph, we add a global max pooling layer right after the last graph convolutional layer. Formally, denote a built drug graph as $\mathcal{G} = (\mathcal{V}, \mathcal{E})$, where $\mathcal{V} \in \mathbb{R}^{N \times F}$ is the set of N nodes each represented by a F-dimensional vector (in this case, $F = 78$) and $\mathcal{E}$ is the set of edges represented as an adjacency matrix $A \in \mathbb{R}^{N \times N}$. The GCN layer is defined by [14] as

$$Z = \tilde{D}^{-\frac{1}{2}}\tilde{A}\tilde{D}^{-\frac{1}{2}}X\Theta \tag{3}$$

where $Z \in \mathbb{R}^{N \times F}$ is convolved feature matrix $\tilde{A}$ is graph adjacency matrix with added self loop and $\tilde{D}$ is the graph diagonal degree matrix, $\Theta \in \mathbb{R}^{N \times C}$ is the trainable parameter matrix.

In out GCN-based model, we make use of three consecutive GCN layers each activated by a ReLU function. Then a global max pooling layer is added to aggregate the whole graph representation.

GAT-Based Graph Representation Learning. We investigate graph attention network (GAT) [29] in our model. Unlike graph convolution techniques, this method proposes an attention-based architecture to learn hidden representations of nodes in a graph by applying a self-attention mechanism. The building block of a GAT architecture is a *graph attention layer*. The GAT layers takes the set of nodes of a graph as input, applies a linear transformation to every node by a weigh matrix $\mathbf{W} \in \mathbb{R}^{F' \times F}$ where F, F' are feature dimensions of input and output nodes, respectively. At the input node i in the graph, the *attention coefficients* between i and its first-order neighbors are computed as

$$e_{ij} = a(\mathbf{W}h_i, \mathbf{W}h_j) \tag{4}$$

The value of e_{ij} indicates the importance of node j to node i. These *attention coefficients* are then normalized by applying a softmax function, then used to compute the layer output as

$$h'_i = \sigma(\sum_{j \in \mathcal{N}_i} \alpha_{ij} \mathbf{W}h_j) \tag{5}$$

where $\alpha(.)$ is a non-linear activation function, and α_{ij} are the normalized *attention coefficients*.

In our model, the GAT-based graph learning architecture includes two GAT layers, activated by a ReLU function, then followed a global max pooling layer to obtain graph representation vector. In details, for the first GAT layer *multi-head-attentions* is applied with number of heads set to 10, number of output features are set identical with the number of input features, 78. The output features of the second GAT is set to 128.

GAT-GCN Combined Graph Neural Network. We consider a combination of GAT and GCN for learning on graphs in proposed model. In detail, the graph neural network starts by a GAT layer which takes graphs as input and returns convolved feature matrix to a GCN layer. Each layer is activated by a ReLu function. The graph representation vector is then computed by concatenating the global max pooling and global mean pooling layers from GCN layer output.

Graph Isomorphism Network. We integrate a recently proposed graph learning method, namely Graph Isomorphism Network (GIN) [31]. This model iş a generalization of Weisfeiler-Lehman (WL) graph isomorphism test, and is theory proven that it achieves maximum discriminative power among graph neural

networks [31]. Specifically, GIN uses a multi-layer perceptrons (MLP) to update
the node features as

$$\mathbf{x}'_i = MLP((1 + \epsilon)\mathbf{x}_i + \sum_{j \in \mathcal{N}(i)} \mathbf{x}_j)$$ (6)

where ϵ is either a learnable parameter or a fixed scalar.

In our model, the GIN-based graph neural net consists of five GIN layers,
each followed by a batch normalization layer. Finally, a global pooling layer is
added to aggregate graph representation vector.

4   Experiments

We conduct an across-county prediction task to estimate health indices of the
US counties, where the BRFSS reports are used as ground truth. Technically,
this is a regression task where, given a county, input is a feature vector extracted
from that county and output is a population health index. In our study, three
primary health indices evaluated include "*poor or fair health*", "*poor physical
health days*", and "*poor mental health days*", referred to as "*generic health*",
"*physical health*", and "*mental health*", respectively.

We employed a linear regression model for this case study. Note that the
same model was also used in previous works [7,18,28]. Specifically, the health
index y^P of a population P can be estimated as follows

$$y^P = \mathbf{w}^\top \mathbf{x}^P + e$$ (7)

where $\mathbf{x}^P$ is the input feature vector extracted from population P and $e \sim
N\left(0, \sigma^2\right)$ is a Gaussian error term.

In our case, the feature vector $\mathbf{x}^P$ can be replaced by the graph-based representation $\mathbf{h}^P$. The weight vector $\mathbf{w}$ can be learned directly from training data.
The dataset is randomly split into set of 2,255 counties (equivalent to 70%) in
every year for training and the remainder (966 counties) for testing. Estimation
accuracy is measured using the Spearman's rank correlation coefficient (rho)
and the root mean squared error (RMSE) between the estimated indices and
the actual health indices (i.e., the ground truth).

4.1   Dataset

We crawled 1,129,928,183 tweets made in years 2014, 2015, and 2016, and geotagged with US latitude/longitude coordinates, which were mapped to US geocodes. The collected geo-tagged tweets were associated with the US counties by
mapping their geo-codes to the Federal Information Processing System (FIPS)
codes using the cartographic boundary files provided by the US Census Bureau
in 2013. There were 3,221 different geo-codes corresponding to 3,221 counties in
the US. Note that we used only tweets with associated latitude/longitude coordinates, those with self-reported location information but without coordinates
were not considered in our study.

We use the BRFSS survey reports as the ground truth for our experiments. The surveys were conducted by the CDC via telephone and collected data from US residents regarding to heir health-related risk behaviors, chronic health conditions, and health outcomes. BRFSS contains more than 450,000 interviews completed each year[3] and is currently the largest health survey system, not only in the US but also in the world. The questionnaires in BRFSS surveys are categorized into core sections including current health status, number of healthy days, inadequate sleep, chronic health conditions, and optional modules such as healthcare access or social context.

We build prediction models estimating three primary health indices corresponding to those in the BRFSS reports. The health indices include (i) poor or fair health - percent of adults that report fair or poor health, referred to as *generic*, (ii) poor physical health days - the average number of reported physically unhealthy days per month, referred to as *physical*, and (iii) poor mental health days - the average number of reported mentally unhealthy days per month, referred to as *mental*. The values of these indices vary in range of 8–40 for generic, 2.4–5.8 for physical, and 2.2–6.3 for mental index.

4.2   Computing Paradigm

To conduct large-scale data processing tasks, including data aggregation, county mapping, linguistic feature extraction (from billions of tweets), and graph construction, we employed Spark on top of Hadoop [32] for our experiments. Spark is a cluster computing platform which enables distributed and parallel computations on a cluster scaled up to 8,000 nodes. Furthermore, Spark is an in-memory based system which keeps data in memory for convenient subsequent iterations, thus allows much faster computations than disk-based systems like Hadoop MapReduce [11]. Specifically, Spark Hadoop cluster comprises 8 CentOS

Table 1. Prediction performance of health indices of the US counties from Twitter data (Pearson's correlation coefficient, larger is better). LIWC is used as linguistic features. Best performance is in **bold**.

Method	2014			2015			2016		
	Generic	Mental	Physical	Generic	Mental	Physical	Generic	Mental	Physical
Agg [7]	0.64	0.51	0.54	0.57	0.49	0.49	0.52	0.46	0.46
Poly [17]	0.18	0.16	0.16	0.16	0.13	0.12	0.15	0.10	0.10
RBF [17]	0.14	0.14	0.12	0.13	0.09	0.10	0.11	0.11	0.13
Avg [16]	0.63	0.53	0.56	0.39	0.43	0.30	0.52	0.45	0.44
Cov [16]	0.15	0.18	0.16	0.12	0.10	0.09	0.15	0.12	0.13
Tang [16]	0.19	0.16	0.17	0.16	0.13	0.12	0.14	0.12	0.13
GNN	**0.69**	**0.57**	**0.62**	**0.61**	**0.52**	**0.51**	**0.54**	**0.51**	**0.47**

---

[3] https://www.cdc.gov/brfss/index.html.

7.2 physical machines, each equipped with Intel® Xeon® E5-26700 (8 cores, 16 threads) CPU, 128 GB RAM, Intel Xeon Phi Coprocessor (60 cores), and 24TB HDD

4.3 Results and Discussion

Table 1 shows performance of our proposed model. We also re-implemented the existing methods for comparison. The Aggregation method (referred to as Agg) [7], considered as the baseline, simply extracts linguistic features from concatenated data. The kernel-based features (Poly and RBF) [17] using kernel method for extracting population representation. The statistical features (Avg, Cov and Tang) [16] capture the statistical relationship between low-level linguistic features.

As shown in the Table 1, the results of our proposed method show a significant correlation with the ground truth. Best performance was obtained in prediction of indices of 2014 ($\rho = 0.69$ in predicting generic health index). The prediction performance then drops steadily over year 2015, and 2016. The degradation in prediction accuracy might probably due to the continual descent of number of tweets collected over the years from 2014 to 2016.

Table 2. Prediction performance of health indices of the US counties from Twitter data (in RMSE, smaller is better). LIWC is used as linguistic features. Best performance is in **bold**.

Method	2014			2015			2016		
	Generic	Mental	Physical	Generic	Mental	Physical	Generic	Mental	Physical
Agg [7]	3.87	0.54	0.63	4.11	0.50	0.65	4.02	0.54	**0.63**
Poly [17]	16.46	1.88	2.54	20.27	2.71	3.44	20.11	2.87	3.28
RBF [17]	21.75	2.02	3.18	24.39	3.47	4.30	23.48	3.15	3.29
Avg [16]	3.89	0.53	0.62	5.25	0.53	0.85	4.04	0.55	0.64
Cov [16]	19.08	1.97	2.77	21.21	2.67	3.51	19.88	2.86	3.15
Tang [16]	16.38	1.90	2.51	20.88	2.84	3.56	20.67	2.99	3.25
GNN	**3.67**	**0.51**	**0.59**	**4.01**	**0.49**	0.64	4.00	**0.53**	0.63

In comparison against the existing methods, our proposed model obtains the best performance in both Pearson's correlation coefficient (Table 1) and RMSE (Table 2). On average, the graph-based model obtains a correlation coefficient of 0.56 on predicting three health indices over three years, compared to 0.52 of the runner-up – the baseline method [7]. The average features [16] also appear to be predictive (average $\rho = 0.47$) for health prediction, while the kernel-based features and other statistical features show poor performance.

Among the graph neural network models that were investigated in our experiments, the GAT model [29] shows best prediction performance. This might probably due to the characteristic of the GAT model that enable specifying different weighs to different nodes (i.e., linguistic features in our case). These results also confirm our idea that the interactions between features are meaningful. The results reported of our proposed method in Tables 1 and 2 are from the GAT model. The combined GAT-GCN and GCN models also shows significant results, for instance, it obtains ρ values of 0.64 and 0.62 in prediction of generic index of 2014, respectively, compared to 0.69 of the best model (GAT). On the contrary, the GIN model does not learn the graph-based population features effectively.

5   Conclusion

In this paper, we introduce a novel method for population health analysis through social media, based on a graph neural network approach. Specifically, we propose to model the interactions in social media using graphs at feature-level then applying a deep graph learning algorithm to train population health prediction model. The proposed approach was evaluated and compared with existing approaches on the task of estimating the primary health indices of the US counties over three years 2014, 2015 and 2016 on Twitter dataset. Experimental results favorably showed the superiority of the proposed approach over existing ones. This also confirms the contribution of the interactions between linguistic features to the task of population health computing. Additionally, these significant results suggest the potential of the approach for analysis tasks at population scale.

References

1. Andalibi, N., Ozturk, P., Forte, A.: Depression-related imagery on Instagram. In: Proceedings of the ACM Conference Companion on Computer Supported Cooperative Work & Social Computing, pp. 231–234 (2015)
2. Atwood, J., Towsley, D.: Diffusion-convolutional neural networks. In: Proceedings of the Advances in Neural Information Processing Systems, pp. 1993–2001 (2016)
3. Bagroy, S., Kumaraguru, P., De Choudhury, M.: A social media based index of mental well-being in college campuses. In: Proceedings of the CHI Conference on Human factors in Computing Systems, pp. 1634–1646. ACM (2017)
4. Belkin, M., Niyogi, P.: Towards a theoretical foundation for Laplacian-based manifold methods. In: Proceedings of the International Conference on Computational Learning Theory, pp. 486–500 (2005)
5. Bruna, J., Zaremba, W., Szlam, A., LeCun, Y.: Spectral networks and locally connected networks on graphs. arXiv preprint arXiv:1312.6203 (2013)
6. Chen, M.K.: The effect of language on economic behavior: evidence from savings rates, health behaviors, and retirement assets. Am. Econ. Rev. **103**(2), 690–731 (2013)
7. Culotta, A.: Estimating county health statistics with Twitter. In: Proceedings of the SIGCHI Conference on Human Factors in Computing Systems, pp. 1335–1344 (2014)
8. De Choudhury, M., Counts, S., Horvitz, E.: Social media as a measurement tool of depression in populations. In: Proceedings of the Annual ACM Web Science Conference, pp. 47–56 (2013)

9. De Choudhury, M., Gamon, M., Counts, S., Horvitz, E.: Predicting depression via social media. In: Proceedings of the International AAAI Conference on Weblogs and Social Media, pp. 128–137 (2013)
10. Defferrard, M., Bresson, X., Vandergheynst, P.: Convolutional neural networks on graphs with fast localized spectral filtering. In: Proceedings of the Advances in Neural Information Processing Systems, pp. 3844–3852 (2016)
11. Dittrich, J., Quiané-Ruiz, J.-A.: Efficient big data processing in Hadoop MapReduce. Proc. VLDB Endow. 5(12), 2014–2015 (2012)
12. Dredze, M., Paul, M.J.: Natural language processing for health and social media. IEEE Intell. Syst. 29(2), 64–67 (2014)
13. Gottschalk, L.A., Gleser, G.C.: The Measurement of Psychological States Through the Content Analysis of Verbal Behavior. University of California Press, Berkeley (1979)
14. Kipf, T.N., Welling, M.: Semi-supervised classification with graph convolutional networks. In: ICLR (2017)
15. Mowery, D., Bryan, C., Conway, M.: Feature studies to inform the classification of depressive symptoms from Twitter data for population health. arXiv:1701.08229 (2017)
16. Nguyen, T., et al.: Using spatiotemporal distribution of geocoded Twitter data to predict US county-level health indices. Future Gener. Comput. Syst. (2018)
17. Nguyen, T., et al.: Kernel-based features for predicting population health indices from geocoded social media data. Decis. Support Syst. 102, 22–31 (2017)
18. Nguyen, T., et al.: Prediction of population health indices from social media using kernel-based textual and temporal features. In: Proceedings of the International Conference on World Wide Web Companion, pp. 99–107 (2017)
19. Niepert, M., Ahmed, M., Kutzkov, K.: Learning convolutional neural networks for graphs. In: Proceedings of the International Conference on Machine Learning, pp. 2014–2023 (2016)
20. Paul, M.J., Dredze, M.: You are what you tweet: analysing Twitter for public health. In: Processing of the International AAAI Conference on Weblogs and Social Media (2011)
21. Paul, M.J., Dredze, M.: A model for mining public health topics from Twitter. Health 11, 16–6 (2012)
22. Pennebaker, J.W., Beall, S.K.: Confronting a traumatic event: toward an understanding of inhibition and disease. J. Abnorm. Psychol. 95(3), 274 (1986)
23. Pennebaker, J.W., Booth, R.J., Boyd, R.L., Francis, M.E.: Linguistic Inquiry and Word Count: LIWC 2015 [Computer software]. Pennebaker Conglomerates Inc. (2015)
24. Pennebaker, J.W., Francis, M.E., Booth, R.J.: Linguistic inquiry and word count: LIWC 2001. Mahway: Lawrence Erlbaum Associates, vol. 71, no. 2001, p. 2001 (2001)
25. Pennebaker, J.W., Mehl, M.R., Niederhoffer, K.G.: Psychological aspects of natural language use: our words, our selves. Ann. Rev. Psychol. 54(1), 547–577 (2003)
26. Reece, A.G., Danforth, C.M.: Instagram photos reveal predictive markers of depression. EPJ Data Sci. 6(1), 15 (2017)
27. Salathe, M., et al.: Digital epidemiology. PLoS Comput. Biol. 8(7), e1002616 (2012)
28. Schwartz, H.A., et al.: Characterizing geographic variation in well-being using tweets. In: Proceedings of the International AAAI Conference on Weblogs and Social Media, pp. 583–591 (2013)
29. Veličković, P., Cucurull, G., Casanova, A., Lio, P., Bengio, Y., Romero, A.: Graph attention networks. In: ICLR (2018)

30. Von Luxburg, U.: A tutorial on spectral clustering. Stat. Comput. **17**(4), 395–416 (2007)
31. Xu, K., Hu, W., Leskovec, J., Jegelka, S.: How powerful are graph neural networks? In: ICLR (2019)
32. Zaharia, M., et al.: Fast and interactive analytics over Hadoop data with Spark. Usenix Login **37**(4), 45–51 (2012)

Joint Sequential Data Prediction with Multi-stream Stacked LSTM Network

Nguyen Thanh Toan[1(✉)], Orçun Gümüş[2], Nguyen Thanh Tam[2],
Nguyen Quoc Viet Hung[3], Rene Hexel[3], and Jun Jo[3]

[1] Ho Chi Minh City University of Technology, Ho Chi Minh City, Vietnam
nguyentoanit@gmail.com
[2] École Polytechnique Fédérale de Lausanne, Lausanne, Switzerland
[3] Griffith University, Gold Coast, Australia

Abstract. Accurate traffic density estimation is essential for numerous purposes such as transit policy development or forecasting future traffic conditions for navigation. Current developments in machine learning and computer systems bring the transportation industry numerous possibilities to improve their operations using data analyses on traffic flow sensor data. However, even state-of-art algorithms for time series forecasting perform well on some transportation problems, they still fail to solve some critical tasks. In particular, existing traffic flow forecasting methods that are not utilizing causality relations between different data sources are still unsatisfying for many real-world applications. In this paper, we have focused on a new approach named multi-stream learning that uses underlying causality in time series. We evaluate our method in a very detailed synthetic environment that we specially developed to imitate real-world traffic flow dataset. In the end, we assess our multi-stream learning on a historical traffic flow dataset for Thessaloniki, Greece which is published by Hellenic Institute of Transport (HIT). We obtained better results on the short-term forecasts compared the widely-used benchmarks models that use a single time series to forecast the future.

Keywords: Sequential deep learning · Traffic flow prediction · Machine learning

1 Introduction

Accurate traffic flow information is required to improve decision-making processes. By efficiently using traffic flow information, individuals may decrease their daily transportation duration, public transports can improve their services by designing their scheduler more intelligently, or government increase traffic operation efficiency [18].

Accurate forecasting can also be very critical for an individual basis. Suppose you are a tourist in Greece, and you want to travel from your Hotel to a famous Archaeological Museum of Thessaloniki by a car. You want to know the duration of this travel to plan the root of your trip in the city. As you can see from Fig. 1,

© Springer Nature Singapore Pte Ltd. 2019
T. D. Le et al. (Eds.): AusDM 2019, CCIS 1127, pp. 77–90, 2019.
https://doi.org/10.1007/978-981-15-1699-3_7

some possible trip routes between you and your target are highly crowded and request more time to cross then the usual duration. However this map shows a snapshot in the time domain, therefore after some period, states may vary, and the perfect path may be changed. So finding the best route possible not only requires the current knowledge but also future forecasts. Big companies like Google is also using traffic flow forecasts to not only calculate the shortest possible road trips from point A to B but also to estimate the paths with the shortest duration possible.

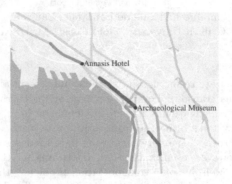

Fig. 1. Primetime traffic flow map snapshot of Thessaloniki, Greece from 3rd January 2018. Colors represent normalized travel times. Red colors represent travel times that are closer to the monthly maximum values, greenish colors represent the travel times that are close to the monthly minimum values. The travel times which are shown as gray represent missing values on the collected data. (Color figure online)

As we explained before traffic flows are not constant with time; they are generally more dense in prime hours of the day and more sparse in others. So estimating the shortest travel duration from point A to B is requiring not only to find a path in the graph but also forecasting future traffic flows of the road network. Many forecasting algorithms which are currently in use are only using a single time series to predict the future values of traffic flows [2]. In this work, we focus on developing a multi-stream deep neural network to obtain forecasts which can learn causality in the multiple time series.

We develop a very detailed synthetic environment to precisely control the variance and causality between different time series. We are inspired by real datasets while implementing the synthetic data generator by trying to include many different scenarios in the simulated environment. For testing our algorithm on real traffic flow data, we use the open data provided by The Hellenic Institute of Transport (H.I.T.) for the city of Thessaloniki, Greece. Using their API, we obtained the travel times of all the routes in Thessaloniki for every 15 min interval from January 2018 to February 2018 [21]. To develop a solution that uses causality in the different time series, we used generative LSTM networks with fusion-learning. Using this approach, we are able to predict future travel times more accurately than other baselines.

Following these contributions, we organise our paper as follow. Section 2 discusses the related work of the time-series forecasting setting. Section 3 formulates the novel setting of multiple time-series prediction. Section 4 presents our approach and model details. All experiments are shown in Sect. 5. Section 6 concludes the paper.

2 Background and Related Work

2.1 Forecasting Using Neural Nets

Artificial neural networks can learn high dimensional patterns via a sequence of non-linear transformations of input data; by allowing us to efficiently model of nonlinear functions. Between different deep learning algorithms, recurrent neural networks (RNN) have been regularly used for forecasting purposes. There are multiple recurrent neural network structures for generative purposes. According to usecases, input and output shapes, and dataset proprieties, the most suitable structure is varying, but limited to three main structural skeletons [13]:

1. *one-to-many:* single input with an output sequence (e.g. image captioning takes a photo and produce a sentence as an output).
2. *many-to-one:* input sequence with a single output (e.g. sentiment analysis in which a given comment is classified as a positive or negative sentiment).
3. *many-to-many:* input sequence with an output sequence (e.g. in machine translation, an RNN reads a sentence in English and then outputs a sentence in French [13]).

2.2 State-of-the-Art Sequential Models

For long input sequences, RNN structure has some issues such as vanishing gradient. Long short-term memory (LSTM) networks address such difficulties by designing a neuron model that can learn to bridge minimal time lags in surplus of 1000 discrete time steps by reinforcing constant error flow through constant error carrousels [8]. An LSTM block consists of multiple functions that try to remember the helpful and forget the unnecessary information from inputs. In particular, LSTM "cells" are associated recurrently to each other, changing the usual hidden units of ordinary recurrent networks. The state unit has a linear loop whose weight is regulated by the forget gate. The output of the cell can be shut off by the output gate. All the gating units have a sigmoid nonlinearity [6]. LSTM networks have been found in various domains such as data integration [12,19], recommendation systems [20,23], time-series data analytics [10,16,22], action recognition [15] and sensor data [9,11] as well as have been a state-of-the-art forecasting model [4].

In this paper, we develop a new LSTM architecture for multi-stream learning by modeling causality patterns between multiple time series and leveraging the depth nature of LSTM layers to capture these patterns. Our multi-stream learning scheme is based on an advanced version of LSTM [5].

3 Problem Formulation

Let $T_k = [x_k^0, x_k^1, x_k^2, \ldots, x_k^t]$ is the time series for the k-th road in the trajectory dataset.

Definition 1. *(Forecasting, f) We denote the travel time of road segment k on time step t as x_k^t. Forecasting is the prediction of future values using past values, finding $E(x_k^t)$ using $x_k^{t-i}, i < t$.*

$$E(x_k^t) = f(x_k^{t-1}, x_k^{t-2}, x_k^{t-3} \ldots, x_k^0) \tag{1}$$

Now we consider the problem of forecasting multiple time series at the same time.

Definition 2. *(Multi-Stream Forecasting, g) Suppose that we have k roads. We denote the travel times of all roads segment in the set on time period t as X^t.*

$$X^t = [x_1^t, x_2^t, x_3^t, \ldots, x_k^t] \tag{2}$$

Multi-stream forecasting is the prediction of future values using past values of all timeseries, finding expected values of X^t, using $X^{t-i}, i < t$.

$$E(X^t) = g(X^{t-1}, X^{t-2}, X^{t-3}, \ldots, X^0) \tag{3}$$

4 Approach

Time series forecasting becomes a harder problem when there are multiple correlated time series to predict. We try to introduce a multi-stream learning method to capture the relations between different time series. Whereas other methods only focus on one-time series forecasting to the future of that time series, our method is tried to capture all intra-time series patterns.

4.1 Overview

We have implemented many-to-one and many-to-many approaches in order to compare the performance. In "many-to-one" scenario we have trained the model by giving all previous data in a window w and requesting the next value after the window. So our model try to predict the output X^t, given a sequence of w inputs $[X^{t-w}, X^{t-w+1}, \ldots, X^{t-1}]$.

In many to many scenario, we used all previous outputs during training. During the training of neural network, we have used first-order gradient-based optimisation of stochastic objective functions, based on adaptive estimates of lower-order moments (ADAM), and we have used backpropagation to find optimal parameters for the model weights [14].

We supposed that traffic density on different streets in a particular city is correlated. So knowing a traffic density of a street is important information about another street. So we can assume that traffic densities on different roads are not independent. In other words $P(x_i^t, x_j^t) \neq P(x_i^t) \times P(x_j^t), i \neq j$.

Let us assume graph G is a graph as denoted as $G = (V, E)$, where E denotes a correlation between two time series and V indicate the travel time series for a street. We construct this graph using 0.5 for correlation threshold. Then we run the spin-glass community detection algorithm on graph G to find correlated travel times.

4.2 Causality Modeling

Using multiple time series can their prediction accuracy. For example, we can consider that time series that represent traffic density in a road may be heavily correlated with the past of others. This means X^t may be correlated with $[X^{t-w} - X^{t-1}]$. In this formulation, $[X^{t-w} - X^{t-1}]$ represents w step previous value of X^t. So in order to improve prediction accuracy, we may want to use these dependencies that are named as Granger causality. That means if past values of $[X^{t-w} - X^{t-1}]$ helps in predicting future values of X^t we say $[X^{t-w} - X^{t-1}]$ granger causes X^t [3].

Our aim in this formulation is capturing Granger causality relationships (as presented in Fig. 2) between different time series. Let us modify graph G, by changing edges to represent not the correlation for that time, but the correlation for one step lagged version ($X^t \sim X^{t-1}$).

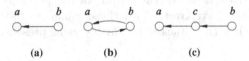

(a)   (b)   (c)

Fig. 2. Causality patterns: (a) direct causality, (b) direct feedback, (c) indirect causality

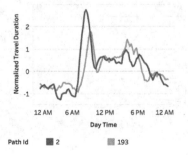

Path Id   ■ 2   ■ 193

Fig. 3. Granger causality between street 193 (Egnatia) and street 2 (Konstantinou Karamanli) on 10/01/2018.

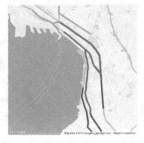

Fig. 4. Detected road communities in Thessaloniki, Greece. Each different color represents a group. Roads in the same communities varying together; in other words, travel times are more correlated. (Color figure online)

For example in Fig. 3, at prime times of 10 January 2018 we first see a peak in street-2 then with a 15-min lag street-193 also becomes crowded. So street-2 can help to predict better street-193. There may be many possibilities that may because that type of lagged causalities; for example, suppose there is a sporting event that finishes at 20:00 at a sports arena close to the street-2. When the event finalised, everybody is leaving the sports arena. So as a result of this we first see a local peak in the street-193 then a step further this may cause other peaks in other streets like street-2. Using this approach, we can detect communities of correlated time series as in Fig. 4.

4.3   Multi-stream Learning

To learn the aforementioned types of causalities and take advantage of lagged correlations, we propose a multi-stream neural network model by stacking LSTM layers. For example, the model for forecasting the time t model is taking all inputs from $t - w$ to $t - 1$ and a generate a point estimation for t. If we have k different roads in our model, the input for our algorithm is a k-dimensional vector at each time step, and we forecast a k-dimensional vector for the next time step.

Existing studies [7] showed deep LSTM architectures can build up a progressively higher level of abstraction of sequence data, and thus, work more effectively. A deep LSTM architecture can be created by stacking multiple LSTMs layers on top of each other. Particularly, the output of the previous layer provides an input sequence for the next layer. With these benefits of stacked LSTMs, we propose our approach with 2 stacked LSTM layers as presented in Fig. 5.

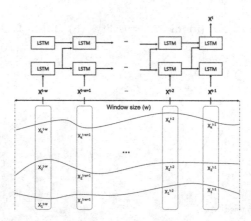

Fig. 5. Multi-stream neural network model; LSTM neuron type with many-to-one structure, $X^t = [x_1^t, x_2^t, \ldots, x_k^t]$. Layers added on top of the other layers.

In general, the two LSTM layers are expected to capture the first-order and second-order causality patterns (Fig. 3). Adding more layers will enable to capture higher-order patterns. However, as we tried with our datasets, two layers are sufficient. This could be explained by the fact that dependencies between more than three time-series (e.g. traffic data) are rarely encountered in practice.

5   Experimental Results

5.1   Setup

Real Dataset. The traffic flow dataset has been collected from CERTH-HIT OpenData Hub BETA web site using their JSON API [21]. The data contains travel times of main streets of Thessaloniki, Greece for each 15 min periods from January to 2018 to February 2018. The data formatted in CSV and contains only the durations in the second format. To visualise roads on the map, we have used another dataset from the same portal which contains shapefile for the paths. The shapefile contains geometry of the streets by multiple points in latitude and longitude format.

Synthetic Dataset. In order to test our forecasting method, we have designed a synthetic data generator. We develop it in such a way that it can imitate the real world scenarios. To do so, we have implemented a very detailed random walk generator that can handle multiple correlated time series. We become able to generate various datasets from a basic random walk to a mixture of many different patterns. Before going deep into the synthetic dataset results, we introduce you the patterns that you can use to generate random walks:

- *Gaussian Random Walk (GRW):* is the sum of a series of independent and identically distributed random variables; x_i taken from a normal distribution with mean equal zero and variance σ^2:

$$x^i = \mathcal{N}(0, \sigma^2); \ X^t = \sum_{i=1}^{t} x^i \qquad (4)$$

- *Correlated Gaussian Random Walk (CGRW):* We expect that our model should be able to learn the causality between different time series. To test it, we are using a correlated time series with a constant lag. To explain it more precisely suppose we have two different Gaussian random walks and x_1^i and x_2^i are ith steps of first and second time series respectively. x_1 and x_2 are the vectors from one multivariate normal distribution for x, so $x = (x_1, x_2)^T$.

$$\begin{pmatrix} x_1 \\ x_2 \end{pmatrix} \sim N \left[\begin{pmatrix} 0 \\ 0 \end{pmatrix}, \begin{pmatrix} 1 & c \\ c & 1 \end{pmatrix} \right] \qquad (5)$$

Then we are using this multivariate normal distribution to create two time-series with a lagged correlation to represent causality. So if we suppose lag constant as l the new timeseries X_1^t and X_2^t will be:

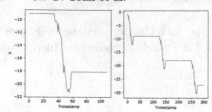
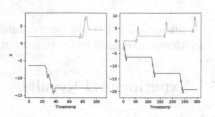

Fig. 6. The left figure shows an example of a short random walk, while the right figure demonstrates how we use this random walk on a time series. To illustration, we use the constant inter-gaps between short walks, but in reality, the gaps can be normally distributed.

Fig. 7. The left figure shows an example of two short random walks with an intra-gap. The right figure demonstrates how we use this pair on two time series.

$$X_1^t = \sum_{i=1}^{t} x_1^i \qquad (6)$$

$$X_2^t = \begin{cases} \sum_{i=1}^{t-l} x_2^i, & \text{if } t > l. \\ 0, & \text{otherwise.} \end{cases} \qquad (7)$$

As you can see from Eq. 7, there is a lag constant l which makes possible predict better X_2^t using X_1^{t-l}.

– *Patterns as Short Random Walks:* There are repetitive patterns in time series of Traffic Flow Data [17]. To imitate this structure, we have used a short random walk (as described in Fig. 6) to represent repetitive patterns along with the time series. The short random walks can be thought of as a similar version of wavelet in a wave.

We have also come up with an idea of using patterns across different series to imitate causality between time series (Fig. 7). For example, if a specific pattern occurs on X_1^t, we may assume that another specific pattern will occur on X_2^{t+l}. We have used this approach on top of correlated random walks to generate causality between series.

Metrics. We use the most common error measurements in time-series forecasting setting:

– *MAE (Mean Absolute Error):* Metric which measures the average magnitude of the errors in a set of predictions, without examining their direction [1].

$$\text{MAE} = \frac{1}{n} \sum_{t=1}^{n} |e_t| \qquad (8)$$

– *Normalized MSE (Mean Squared Error):* Normalized (Root) Mean Squared Error which facilitates the comparison between models with different scales [1].

$$\text{NMSE} = \frac{\sqrt{\frac{1}{n}\sum_{t=1}^{n} e_t^2}}{max(Y) - min(Y)} \tag{9}$$

The reason that we have selected NMSE over MSE is that our two datasets have different scales and need to be compared within the same normalized scale.

- *The coefficient of determination (R^2):* is the fraction of the variance in the dependent variable that is predictable from the independent variable. R^2 is also not scaling sensitive. However, R^2 is not used for measuring forecast accuracy, but rather a secondary indicator, as it does not account for overfitting and bias [1].

Baselines. Different traditional and state-of-the-art forecasting methods are used.

- *Moving Average (MA):* is one of the most popular forecasting techniques, taking the average of the last n values as the future prediction [1].

$$f_{MA(n)}(x_i^{t-1}, x_i^{t-2}, \dots, x_i^0) = \frac{1}{n} \times \sum_{k=1}^{n} x_i^{t-k} \tag{10}$$

- *Exponential smoothing (ES):* is another popular technique. Whereas in MA, the past observations are weighted equally, ES uses exponential functions to assign exponentially decreasing weights over time, with a smoothing constant a [1].

$$f_{ES(a)} = a(x_i^{t-1} + (1-a)x_i^{t-2} + (1-a)^2 x_i^{t-3} \dots) + (1-a)^{t-1} x_i^0 \tag{11}$$

- *Double exponential smoothing (DES):* is a technique to forecast time series with a trend. Whereas ES only processes the levels in time series, DES also handles the slopes of time series by exponentially smoothing it [1].

$$f_{2ES(a,b)} = s^{t-1} + b^{t-1} \tag{12}$$

- *LSTM:* is a state-of-the-art deep learning based forecasting model, whose detailed structured can be found in [4,5].

We have compared different neural network models in varying structures. All these implementations will be discussed with positive and negative outcomes in the results section. We use 3-fold validation to avoid overfitting.

5.2 Experiments on Traffic Flow Data

As explained earlier, by implementing multi-stream learning approach we use the information of multiple time series to forecast one, in order to improve the forecasting performance as compared to the case when we only use one time series. The results on traffic flow data are shown in Table 1.

As you can see from Table 1, multi-stream learning model (J-LSTM) has slightly better error measurements. Even the model J-LSTM uses exactly the same neural network structure with the LSTM model (same gates, same layer count, same neuron count, same activation function, etc) it has achieved better results.

Table 1. Comparison of forecasting methods on real dataset

Methods	R^2	MAE	Normalized MSE
MA	0.6910	805.8124	0.0828
ES	0.7091	758.6089	0.0803
DES	0.6867	816.9235	0.0833
LSTM	0.7308	702.0266	0.0772
J-LSTM	**0.7403**	**678.5740**	**0.0759**

Table 2. Comparison of forecasting methods on synthetic dataset

Methods	R^2	MAE	Normalized MSE
MA	0.9986	1.7575	0.0122
ES	0.9983	2.2013	0.0136
DES	0.9988	1.5312	0.0114
LSTM	0.9989	1.3339	0.0106
J-LSTM	**0.9991**	**1.0884**	**0.0096**

5.3 Experiments on Simulations

The above results on real data showed that multi-stream neural network model is capable to learn causality relationship between different time series. In this section, we want to explore if the same model is still robust to forecasting future values of different time-series datasets. To this end, we evaluate our approach on the simulated datasets created by the aforementioned random-walk mechanisms.

In our experiments, we have used $l = 1$ to generate correlated random walks. The activation function of the neural network is set to 0.0001 and data is normalized between 0 and 1. We have generated simulated datasets which contain correlated time series to test our algorithm. The dataset only contains 2 different time series with 0.5 correlation with one step offset on top of different trends and seasonal patterns. To compare the performance of our algorithm we use three different error measure MAE, R^2, and Normalized MSE.

The results of synthetic datasets have been shared in Table 2. Among all methods, our implemented method, J-LSTM perform best compared with others. All parameters for forecasting methods have been determined using grid search for best performance on normalised mean square error. Exponential smoothing applied with $a = 0.8$, double exponential smoothing applied with $a = 0.8, b = 0.5$, moving average method has been applied using $w = 1$.

As demonstrated, the simulated dataset has a sinusoidal and polynomial signal. This sinusoidal and polynomial signal makes it perfect to be learned by neural networks. As a result of this, neural network models such as LSTM perform better than usual forecasting methods like ES or DES. However even standard neural networks models can learn trend, and sinusoidal signal they are not able to catch the correlation between different time series. Our multi-stream learning is also catching correlation information, and error becomes diminutive compared to other forecasting algorithms.

5.4   Qualitative Showcases

Correlation. For the sake of simplicity, we have focused on two roads which are highly correlated. To find this two-roads forecasting, we have shifted travel times with a one-time step and calculate the pairwise correlations then we have selected the pair with the highest correlation. The maximum pairwise correlation that we have calculated is 0.65. The street pair which have a maximum correlation (193 and 2) are used in our experiments. Figure 8 shows that multi-stream learning is capable of learning casual relationships which fall into more than two series.

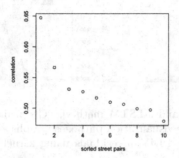

Fig. 8. The travel times of road pairs sorted by their correlation. Visualized only ten road pairs which have maximum pairwise correlation.

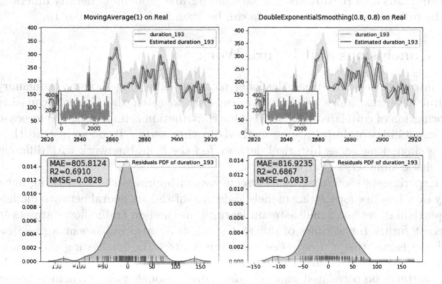

Fig. 9. Results of MA and DES. Green lines represent estimations, blue lines are real vales. For visualization purposes, graphs are focused on the last 100 steps. We also visualize the PDFs of residuals using kernel as Gaussian distribution. (Color figure online)

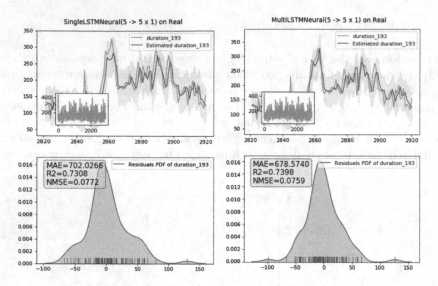

Fig. 10. Results of LSTM and J-LSTM methods. Green lines represent estimations, blue lines are real vales. For visualization purposes, graphs are focused on the last 100 steps. We also visualize the PDFs of residuals using kernel as Gaussian distribution. (Color figure online)

Predictions and Residuals. The estimations and probability density functions of the residuals of different methods can be examined in Figs. 9 and 10.

6   Conclusions and Future Work

We introduce a multi-stream learning method using long-short-term memory neural networks. More precisely, we are able to catch the Granger causality information of correlated time series. Our contribution is improving the forecasting capabilities of deep learning methods in time series. To do so instead than using single time series historical data we fed the neural network with different correlated time series.

Experiments show that multi-stream model improves the forecasting accuracy over baseline forecasting models and state-of-the-art neural network models. In particular, we use a multi-stream learning method on traffic flow datasets to forecast future travel times of different streets. We also implement a practical and precise synthetic time series generator to test the forecasting abilities of different methods. We are able to generate time series which contains consistent patterns on correlated walks or sinusoidal seasonality or polynomial trends or even a mixture of all of them. Using controlled experiments on the generated dataset, we become sure that multi-stream learning method can learn the Granger causality between different time series. For future work, we will fulfill the limitation of this work by exploring further evaluations such as the Granger causality relationships among time series.

References

1. Armstrong, J.S.: Evaluating forecasting methods. In: Armstrong, J.S. (ed.) Principles of Forecasting. ISOR, vol. 30, pp. 443–472. Springer, Boston (2001). https://doi.org/10.1007/978-0-306-47630-3_20
2. Benkachcha, S., Benhra, J., El Hassani, H.: Causal method and time series forecasting model based on artificial neural network. Int. J. Comput. Appl. **75**(7), 37–42 (2013)
3. Dahlhaus, R., Eichler, M.: Causality and graphical models in time series analysis. In: Oxford Statistical Science Series, vol. 27, January 2003
4. Fawaz, H.I., Forestier, G., Weber, J., Idoumghar, L., Muller, P.A.: Deep learning for time series classification: a review. Data Min. Knowl. Disc. **33**(4), 917–963 (2019)
5. Ghaderi, A., Sanandaji, B.M., Ghaderi, F.: Deep forecast: deep learning-based spatio-temporal forecasting. arXiv preprint arXiv:1707.08110 (2017)
6. Goodfellow, I., Bengio, Y., Courville, A.: Deep Learning. MIT Press, Cambridge (2016)
7. Graves, A., Jaitly, N., Mohamed, A.R.: Hybrid speech recognition with deep bidirectional LSTM. In: ASRU, pp. 273–278 (2013)
8. Hochreiter, S., Schmidhuber, J.: Long short-term memory. Neural Comput. **9**(8), 1735–1780 (1997)
9. Huggard, H., Koh, Y.S., Riddle, P., Olivares, G.: Predicting air quality from low-cost sensor measurements. In: Islam, R., et al. (eds.) AusDM 2018. CCIS, vol. 996, pp. 94–106. Springer, Singapore (2019). https://doi.org/10.1007/978-981-13-6661-1_8
10. Hung, N.Q.V., Anh, D.T.: Combining SAX and piecewise linear approximation to improve similarity search on financial time series. In: ISITC, pp. 58–62 (2007)
11. Hung, N.Q.V., Jeung, H., Aberer, K.: An evaluation of model-based approaches to sensor data compression. TKDE **25**, 2434–2447 (2013)
12. Jo, J., Tsunoda, Y., Stantic, B., Liew, A.W.-C.: A likelihood-based data fusion model for the integration of multiple sensor data: a case study with vision and lidar sensors. In: Kim, J.-H., Karray, F., Jo, J., Sincak, P., Myung, H. (eds.) Robot Intelligence Technology and Applications 4. AISC, vol. 447, pp. 489–500. Springer, Cham (2017). https://doi.org/10.1007/978-3-319-31293-4_39
13. Karpathy, A.: The unreasonable effectiveness of recurrent neural networks. Andrej Karpathy Blog, vol. 21 (2015)
14. Kingma, D.P., Ba, J.: Adam: a method for stochastic optimization. arXiv preprint arXiv:1412.6980 (2014)
15. Liang, B., Zheng, L., Li, X.: Sequential deep learning for action recognition with synthetic multi-view data from depth maps. In: Islam, R., et al. (eds.) AusDM 2018. CCIS, vol. 996, pp. 360–371. Springer, Singapore (2019). https://doi.org/10.1007/978-981-13-6661-1_28
16. Liang, H., Du, H., Wang, Q., et al.: Real-time collaborative filtering recommender systems. In: AusDM, pp. 227–231 (2014)
17. Lv, Y., Duan, Y., Kang, W., Li, Z., Wang, F.Y.: Traffic flow prediction with big data: a deep learning approach. TITS **16**(2), 865–873 (2014)
18. Van Hinsbergen, C.P.I, Lint, J., Sanders, F.M.: Short term traffic prediction models. In: World Congress on Intelligent Transport Systems, vol. 7, November 2007
19. Tam, N.T., Hung, N.Q.V., Weidlich, M., Aberer, K.: Result selection and summarization for web table search. In: ICDE, pp. 231–242 (2015)

20. Tang, X., Xu, Y., Abdel-Hafez, A., Shlomo, G.: A multidimensional collaborative filtering fusion approach with dimensionality reduction. In: AusDM (2014)
21. Hellenic Institute of Transport, H.I.T.: Traffic flow (2018). H.I.T. Portal http://opendata.imet.gr/dataset
22. Wee, C.K., Nayak, R.: An approach to compress and represents time series data and its application in electric power utilities. In: Islam, R., et al. (eds.) AusDM 2018. CCIS, vol. 996, pp. 107–120. Springer, Singapore (2019). https://doi.org/10.1007/978-981-13-6661-1_9
23. Yin, H., Chen, L., Wang, W., Du, X., Hung, N.Q.V., Zhou, X.: Mobi-SAGE: a sparse additive generative model for mobile app recommendation. In: ICDE, pp. 75–78 (2017)

Show Me Your Friends and I'll Tell You Who You Are. Finding Anomalous Time Series by Conspicuous Cluster Transitions

Martha Tatusch(✉) [iD], Gerhard Klassen [iD], Marcus Bravidor [iD], and Stefan Conrad [iD]

Heinrich Heine University, Universitätsstr. 1, 40225 Düsseldorf, Germany
{tatusch,klassen,bravidor,conrad}@hhu.de

Abstract. The analysis of time series is an important field of research in data mining. This includes different sub areas like trend analysis, outlier detection, forecasting or simply the comparison of multiple time series. Clustering is also an equally important and vast field in time series analysis. Different clustering algorithms provide different analysis aspects like the detection of classes or outliers. There are various approaches how to apply cluster algorithms to time series. Previous work either extracted subsequences or feature sets as an input for cluster algorithms. A rarely used but important approach in clustering of time series is the grouping of data points per point in time. Based on this technique we present a method which analyses the transitions of time series between clusters over time. We evaluate our approach on multiple multivariate time series of different data sets. We discover conspicuous behaviors in relation to groups of sequences and provide a robust outlier detection algorithm.

Keywords: Outlier detection · Time series analysis · Clustering

1 Introduction

Time series data is collected in various domains. Not only the behavior of users on different platforms, but also the tracking of vehicles and objects or the recording of financial or weather data can be displayed as time series. For further analysis, the various data types can be converted into numerical (mostly discrete) values so that sequences of numerical vectors are derived. These can then be processed in a variety of ways. Information can be obtained through analyses such as clustering, prediction or comparison of time series and different outlier detection methods.

Depending on the context, different aspects can be relevant for the user. For example, not all clustering algorithms consider the same types of clusters, and outlier detection techniques do not always address the same types of outliers. In some cases, very special solutions have to be found for specific problems, whereby there are many algorithms that can be applied to a wide range of application areas.

© Springer Nature Singapore Pte Ltd. 2019
T. D. Le et al. (Eds.): AusDM 2019, CCIS 1127, pp. 91–103, 2019.
https://doi.org/10.1007/978-981-15-1699-3_8

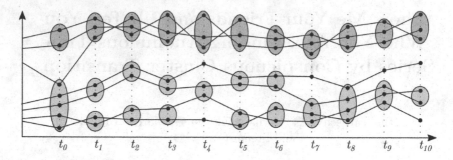

Fig. 1. Example for a time series over-time clustering. The blue color indicates stable clusters while red stands for instability. (Color figure online)

In this paper we focus on databases of multivariate time series with discrete values, same length and equivalent time steps. We detect anomalous subsequences with regard to groups of time series of the given database. Therefore we cluster the multivariate data of all time series per timestamp and analyze the stability of all subsequences over time. Thereby we call the resulting clustering *over-time clustering*. In Fig. 1 an example for such a clustering is displayed. For the sake of simplicity, only univariate time series are plotted. Since the data is clustered independently at each point in time, there is at first no time-related connection between the clusterings.

There are several proposals for clustering time series depending on the application. Some methods cluster the time series of a database as a whole [10,12,19], extract feature sets first [22], or consider subsequences of a single time series only [3]. However, these are not suitable when it comes to detecting irregularities or gathering information per time point.

Outlier detection in time series is in most cases not based on clustering. Because of various underlying data such as single or multiple time series with uni- or multivariate data points and different definitions of what an outlier is, there are several approaches to their identification. Some papers consider data points [1] or subsequences [15] that are anomalous with regard to a single time series [5,17], such as peaks. Others look for so called *change points* [6,16], that imply that the course of the considered time series significantly changes from that point on. Yet others analyse data from several time series that are very similar, such as sensor data, and detect irregularities in relation to the entire data set [1,11,13]. Finding these abnormalities usually presupposes that either the course of a single time series follows consistent patterns or that the courses of several time series are highly correlated.

In this paper we assume that the exact course of the individual time series is not important, but the trend which groups of sequences follow. By anomalies we denote subsequences that deviate from one trend and therefore cannot be assigned steadily to a group of sequences. In that case, we say that the sequence possesses a weak stability. We present an algorithm that identifies such unstable sequences in a database of multivariate time series and is robust against missing data points.

2   Related Work

Anomaly detection in time series is a wide field of research. It can be distinguished in the detection of outliers within a single time series and the detection of outliers in multiple time series. Outliers in single time series are usually categorized in two classes:

Additive outliers, which represent surprisingly large or small values in a short period. In case additive outliers occur consecutively they are often summarized as additive outlier patches.

Innovational outliers are characterized by their impact on subsequent observations. Additionally the influence of innovational outliers can grow with time.

There are also several different categories of outliers, which can be described as a mix of both main classes. For example, additive outliers which cause a move of following observations to a new level are called *level shift outliers* and have a permanent impact on the ongoing time series. In case the influence of the level shift outlier is decreasing over time, it is called a *transient change outlier*. Additive outliers, which occur periodically are named *seasonal additive outliers*.

Additive and innovational outliers are often identified with extensions of autoregressive-moving-average (ARMA) models [2,18]. Other techniques include the use of decomposition methods such as STL, a seasonal-trend decomposition procedure based on LOESS [7]. Yet other methods evaluate derivatives of the dynamic time warping (DTW) [20] similarity in order to detect anomalies.

The detection of outliers in multiple time series is handled differently. Methods of this kind are often using the peers of a time series to determine whether it is anomalous or not. Beside other techniques, recent approaches use Probabilistic Suffix Trees (PST) [21] and Random Block Coordinate Descents (RBCD) [23] in order to detect outliers. However, while these approaches focus on the deviation of one time series to the others, we focus on the behaviour of a time series compared to its peers. More concretely, we assume that a time series which has a similar development to a group of other time series over a subsequence is expected to move on with the same group. Therefore we first cluster per point in time and then analyse the transition of time series regarding these clusters. This is realized by the analysis of cluster transitions of time series over time. Transitions of this kind are also analysed in cluster evolution methods. Landauer et al. [14] makes use of such a method in order to calculate an anomaly score for a single time series in a sliding window. In contrary to Landauer et al. we relate to multiple time series. The analysis of the time series behavior not only reveals new kinds of outliers but also detects different types of additive and innovational outliers.

This approach is very different from clustering whole time series or their subsequences, since the outlier detection would rely on the single fact whether a sequence is assigned to a cluster or not. Such an approach would not take the cluster transitions of the time series into account, which can be an expressive feature on its own. Hence, our approach detects anomalous subsequences, although they would be assigned to a cluster in a subsequence clustering.

3   Fundamentals

In order to create a good basis of knowledge to avoid later misunderstandings, we will provide some definitions which our work is based on. As these terms are used in many different areas, it is useful to explain which interpretations are considered in this paper.

Definition 1 (Time Series). *A multivariate time series* $T = o_{t_1}, \ldots, o_{t_n}$ *is an ordered set of* n *real valued data points of arbitrary dimension. The data points are chronologically ordered by their time of recording, with* t_1 *and* t_n *indicating the first and the last timestamp, respectively.*

Definition 2 (Data Set). *A data set* $D = T_1, \ldots, T_m$ *is a set of* m *time series of same length and equal points in time. The set of data points of all time series at a timestamp* t_i *is denoted as* O_{t_i}.

Definition 3 (Subsequence). *A subsequence* $T_{t_i,t_j,l} = o_{t_i,l}, \ldots, o_{t_j,l}$ *with* $j > i$ *is an ordered set of successive real valued data points beginning at time* t_i *and ending at* t_j *from time series* T_l.

Definition 4 (Cluster). *A cluster* $C_{t_i,j} \subseteq O_{t_i}$ *at time* t_i, *with* $j \in \{1, \ldots, q\}$ *being a unique identifier (e.g. counter), is a set of similar data points, identified by a cluster algorithm or human. This means that all clusters have distinct labels regardless of time.*

Definition 5 (Cluster Member). *A data point* $o_{t_i,l}$ *at time* t_i, *that is assigned to a cluster* $C_{t_i,j}$ *is called a member of cluster* $C_{t_i,j}$.

Definition 6 (Noise). *A data point* $o_{t_i,l}$ *at time* t_i *is considered as noise, if it is not assigned to any cluster.*

Definition 7 (Clustering). *A clustering is the overall result of a clustering algorithm or the set of all clusters annotated by a human for all timestamps. In concrete it is the set* $\zeta = \{C_{t_1,1}, \ldots, C_{t_n,q}\}$ *of all* q *clusters.*

In Fig. 2 an example for the above definitions can be seen. The data points of a data set containing five time series $(T_a, T_b, T_c, T_d, T_e)$ are clustered for the timestamps t_i, t_j and t_k. For simplicity, all data points of a time series T_l are denoted by the identifier l.

In t_i the data points $o_{t_i,a}, o_{t_i,b}$ of time series T_a and T_b are cluster members of cluster $C_{t_i,l}$. The data point $o_{t_i,e}$ is marked as noise, as it is not assigned to any cluster in t_i. In total, the shown clustering consists of five clusters. It can be described by the set $\zeta = \{C_{t_i,l}, C_{t_i,u}, C_{t_j,v}, C_{t_j,f}, C_{t_k,g}\}$.

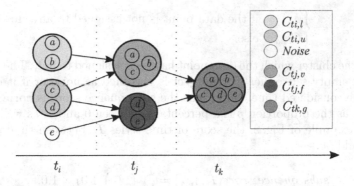

Fig. 2. Example for the transitions of time series $T_a, \ldots, T_e$ between clusters over time.

4   Method

After the clarification of important foundations, the basic idea of the algorithm is described. Therefore further terms have to be explained before.

Let $C_{t_i,a}$ and $C_{t_j,b}$ be two clusters, with $t_i, t_j \in \{t_1, \ldots t_n\}$. First, we introduce the term *temporal cluster intersection* for the purpose of measuring the stability of a time series:

$$\cap_t \{C_{t_i,a}, C_{t_j,b}\} = \{T_l \mid o_{t_i,l} \in C_{t_i,a} \wedge o_{t_j,l} \in C_{t_j,b}\}$$

with $l \in \{1, \ldots, m\}$. The result is the set of time series that are assigned to both of the clusters under consideration. This means all sequences that were grouped together at time t_i and t_j. The transition of a time series from t_i to t_j can now be described by the proportion of cluster members from the corresponding cluster in t_i who migrated together into the cluster in t_j:

$$p(C_{t_i,a}, C_{t_j,b}) = \begin{cases} \emptyset & \text{if } C_{t_i,a} = \emptyset \\ \frac{|C_{t_i,a} \cap_t C_{t_j,b}|}{|C_{t_i,a}|} & \text{else} \end{cases}$$

with $t_i < t_j$. In Fig. 2 an example for transitions of time series between clusters is sketched. There, the proportion for $C_{t_i,l}$ and $C_{t_j,v}$ would be

$$p(C_{t_i,l}, C_{t_j,v}) = \frac{|\{T_a, T_b\}|}{|\{o_{t_i,a}, o_{t_i,b}\}|} = \frac{2}{2} = 1.0.$$

This proportion can be used to measure the stability of a sequence with a *subsequence score*. It is defined as

$$subsequence\_score(T_{t_i,t_j,l}) = \frac{1}{k} \cdot \sum_{v=i}^{j-1} p(cid(o_{t_v,l}), cid(o_{t_j,l}))$$

with $l \in \{1, \ldots, m\}$, $k \in [1, j-i]$ being the number of timestamps between t_i and t_j where the data point exists and cid, the cluster-identity function

$$cid(o_{t_i,l}) = \begin{cases} \emptyset & \text{if the data point is not assigned to any cluster} \\ C_{t_i,a} & \text{else} \end{cases}$$

returning the cluster which the data point has been assigned to in t_i. The function returns an empty set, either if the object is classified as noise or if it does not exist at the considered time. Note, that the subsequence score is normalized to $[0,1]$ by k, as the proportion p is a percentage between 0 and 1, as well.

In the example of Fig. 2, the score of time series T_a between time points t_i and t_k would be:

$$subsequence\_score(T_{t_i,t_k,a}) = \frac{1}{2} \cdot (1.0 + 1.0) = 1.0.$$

A notable characteristic is, that the score is always 0, if the last data point of the considered subsequence is marked as noise. However, this circumstance does not lead to any handicaps in most cases as all partial sequences of these subsequences are treated normally. Nevertheless, the handling of sequences with an endpoint that is labeled as noise will be analyzed in more detail later on.

For now describing the concrete procedure of detecting conspicuous sequences, we first provide a vague definition of them:

Definition 8 (Anomalous Subsequence). *A subsequence $T_{t_i,t_j,l}$ is called anomalous, if it is significantly more unstable than its cluster members at time t_j.*

With the help of the subsequence score which measures the stability of a subsequence, anomalous ones can now be distinguished by comparing the stability of grouped subsequences at a given time point. Every possible subsequence gets an outlier score indicating the probability of being anomalous, by calculating the deviation of its stability from the best subsequence score of its cluster. A formal description of the best subsequence score can be given by:

$$best\_score(t_i, C_{t_j,a}) = max(\{subsequence\_score(T_{t_i,t_j,l}) \mid cid(o_{t_j,l}) = C_{t_j,a}\})$$

The outlier score of a subsequence is then calculated as follows:

$$outlier\_score(T_{t_i,t_j,l}) = best\_score(t_i, cid(o_{t_j,l})) - subsequence\_score(T_{t_i,t_j,l})$$

As the best score lies between 0 and 1, an outlier score of 100% can only be achieved in completely stable clusters. The smaller the best score of the considered cluster is, the smaller is the greatest possible outlier score.

Regarding the example in Fig. 2, the time series T_d would get the following *outlier_score* between time points t_i and t_k:

$$outlier\_score(T_{t_i,t_k,d}) = 1.0 - (0.5 \cdot (0.5 + 1.0)) = 0.25$$

With the outlier score, now a more precise definition of an outlier can be given.

Definition 9 (Outlier). *Given a threshold* $\tau \in [0,1]$, *a subsequence* $T_{t_i,t_j,l}$ *is called an outlier, if its probability of being an outlier is greater than or equal* τ. *That means, if*

$$outlier\_score(T_{t_i,t_j,l}) \geq \tau.$$

Although τ is a constant, it can be interpreted as a dynamic threshold. That is, because the greatest possible deviation from the best subsequence score – and thus the greatest outlier score – depends on the best score of the considered cluster. Clusters with low stability have a lower probability of containing an outlier than stable ones, since all their cluster members show irregularities and that represents a pattern of instability. In this context, the small subsequence score is thus not conspicuous.

Intuitive outliers from the over-time clustering that were marked as noise get a special treatment. Subsequences that consist entirely of noise data points are automatically identified as outliers. Since subsequences whose last data point is labeled as noise are not assigned to a cluster from which the best score can be calculated, no outlier score can be determined for them. Therefore, they are not included in the regular outlier calculation. In the following we will differentiate between *anomalous subsequences*, *intuitive outliers* and *noise*.

Take another look at the case where the last element of an examined subsequence $T_{t_i,t_j,l}$ is marked as noise. Suppose the subsequence $T_{t_i,t_{j-1},l}$ gets a high outlier score and is detected as outlier. Then one would expect that the subsequence under consideration $T_{t_i,t_j,l}$ would be identified as an outlier as well. This will only be the case, if the previous data point was categorized as noise as well and the sequence was therefore recognized as an intuitive outlier. However, for the sequence $T_{t_i,t_k,l}$ with $k > j$, which at the last time point t_k is assigned to a cluster again for the first time this would also be the case. Thus in the end $T_{t_i,t_j,l}$ would be covered.

Yet a marginal case is when a data point is labeled as noise at the last time of the entire time series. In this scenario, a sequence with end time t_m would never be detected as an outlier if it is not marked as noise in t_{m-1}.

Remark 1 (Stability). The stability is not only influenced significantly by a small sample size when considering constant data points [4]. When examining the over-time stability, a small sample size leads to high sensitivity to cluster transitions, as well. As more data points are considered, the simpler it is to draw meaningful conclusions about the stability.

5   Experiments

In order to evaluate the presented method, we performed several experiments on different real world data. We also present two artificially generated data sets which are used to illustrate the handling of some marginal cases. In order to cluster the data per point in time, we used DBSCAN [9] with adapted parameters.

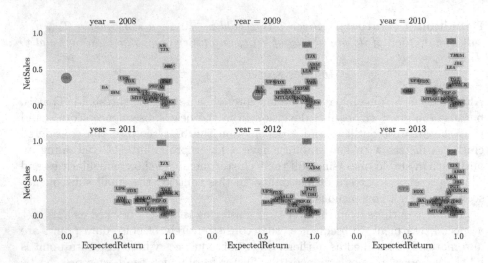

Fig. 3. Two dimensional experiment on the EIKON Financial Data Set with $\tau = 0.6$, $minPts = 2$ and $\epsilon = 0.15$. The colors indicate cluster belongings, whereby grey objects represent outliers. Circles are outliers by distance and boxes are intuitive outliers, as well. Red color or font indicates noise. (Color figure online)

5.1   EIKON Financial Data Set

Eikon is a set of software products released by Refinitiv (formerly Thomson Reuters Financial & Risk). It contains a database with financial data of thousands of companies for the past decades. For illustration reasons we randomly selected thirty companies and two features. The selected features are a figures which were taken from the balance sheet of the company. In economics it is common to normalize these figures by the companies' total assets in order to make it comparable to other companies. Beside this, we normalized the features with a min-max normalization. The clustering was done with DBSCAN and $\epsilon = 0.15$, $minPts = 2$ as parameters. The outlier detection parameter was chosen to be $\tau = 0.6$. In Fig. 3 one can see the illustrated results. The presented technique found two outlier subsequences. The first, which is labeled as *GM* is detected from the year 2008 until 2009. This is because *GM* is noise in the year 2008, which leads to a subsequence score of 0. In 2009 *GM* merges with a cluster, which has a high reference score. The second outlier detected is the subsequence $T_{t_{2009}, t_{2013}, KR}$. It is detected as an intuitive outlier.

5.2   Airline On-Time Performance Data Set

The Airline on-time performance data set [8] was originally collected by the U.S. Department of Transportation's Bureau of Transportation Statistics. It contains records of 3.5 million flights. Every flight has a set of 29 features, such as the departure delay, the delay reason, the arrival delay and the airline which processed the flight. In order to detect anomalies in this data set, we constructed

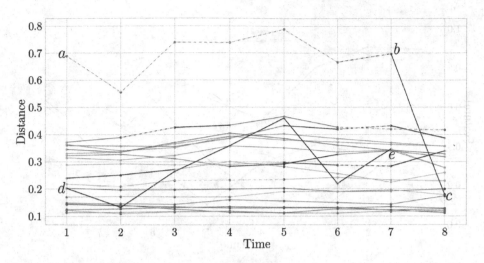

Fig. 4. One dimensional experiment on the Airline On-Time Performance Data Set with $\tau = 0.4$, $minPts = 3$ and $\epsilon = 0.03$. Black sequences represent anomalies, while white dashed ones stand for intuitive outliers. The color of the dots emphasize which cluster the data points are assigned to. Red dots represent noise. (Color figure online)

a time series for every airline by calculating the average of their features for every day. Before applying our technique, we normalized the data with the min-max normalization and clustered it with DBSCAN. Every observation represents a flight of an airline. In order to illustrate the results we executed our algorithm to one feature, namely the flight distance. We applied DBSCAN for eight time points with the following parameters: $minPts = 3$ and $\epsilon = 0.03$. Additionally we chose $\tau = 0.4$. The result can be seen in Fig. 4.

The figure shows two kinds of outliers: Intuitive outliers and outliers which were identified by their distance to a reference time series. Since the time series which is labeled with the points a, b and c has a large distance to other time series it is detected as an intuitive outlier from a to b. Due to this, the time series' accumulated subsequence score is zero and thus it is also detected as an outlier at the last time stamp c. From point a to b it is not detected as an outlier by it's distance to the reference subsequence score, since the neighborhood of the sequence at time point 8 have also a low stability score. Regarding the time points 1 to 8 and the objects in the neighborhood, there are at most two peers which remained together. The subsequence labeled with d and e is a good example for the presented method. It illustrates the detection of outliers by the change of cluster neighbors of the subsequence.

5.3 Simulated Data

In order to test our method in a targeted manner, two experiments were performed on simulated data. Both a univariate and a multivariate data set with two features are considered. In both cases, a time span of 8 time points is examined.

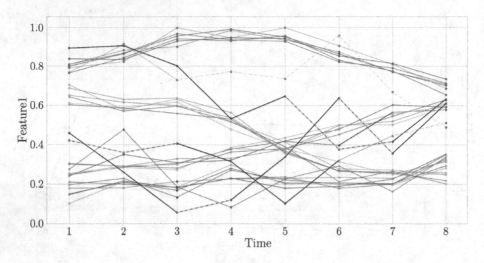

Fig. 5. Illustration of the detected outliers on the simulated one-dimensional data set with $\tau = 0.55$, $minPts = 3$ and $\epsilon = 0.05$. Black sequences represent anomalous subsequences, while white dashed ones stand for intuitive outliers. The color of the dots emphasize which cluster the data points are assigned to. Red dots represent noise. (Color figure online)

The one-dimensional data set was generated so that initially four starting points (for four groups) were selected. In addition, the maximum deviation from the centroid and the number of members were chosen for each group. The centroids were then calculated randomly for each time point, whereby the distance of the centroids of a cluster of two successive time points could not exceed 0.06. After generating the normal data points, 5 outlier sequences were randomly inserted. The starting points were chosen randomly and the distance between two consecutive points could not be greater than 0.3. For all points, care was taken to ensure that they were between 0 and 1.

As shown in Fig. 5, anomalous sequences from five time series have been found. Regarding the first time stamp the first and second black line show time series that are entirely recognized as conspicuous ones. Since their data points often switch between being noise (red dots) and different cluster members, this result is meaningful. Between time point 6 and 7 one additional black line in added. This can be explained by the stability of the sequence's cluster at time 7. All its cluster members migrate together from time point 6 to 7, so that an outlier is very conspicuous.

Looking at the completely randomly generated time series with the uppermost noise point at time 2, it is noticeable that it was not recognized by our algorithm. This is due to the fact that the purple cluster at time 3 and the turquoise cluster at time 5 do not have a high stability and the deviation of the sequence from the best possible score is therefore not very large. In the last time

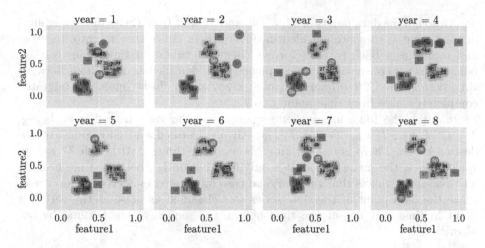

Fig. 6. Illustration of the detected outliers with $\tau = 0.5$, $minPts = 4$ and $\epsilon = 0.11$ on the artificially generated data two-dimensional set. The colors indicate cluster belongings, whereby grey objects represent outliers. Circles are outliers by distance and boxes are intuitive outliers, as well. Red color or font indicates noise. (Color figure online)

points, the time series migrates stably with the yellow cluster, so that it does not behave uncommonly.

If the data points of a time series change from one point in time to another from a cluster to noise, they are not initially interpreted as conspicuous. This is a problem if the time series remains as noise as the time at which it split from the cluster is not recognized as an intuitive outlier. This behavior can for example be seen in the striped line regarding the first time stamp. Between the times 6 and 7, the sequence was not detected as an outlier.

The second data set was created as follows: First, three starting points as centroids and the number of members of the three clusters were chosen. The maximum deviation of two consecutive centroids was set to 0.05 and that of the member data points to the centroid was set to 0.1. One time series was assigned to each group, which was allowed to deviate from the centroid by up to 0.25. Finally, two time series with completely random data points were added, so that a total of 5 outlier sequences should be noticeable. Here, too, we made sure that all data points are between 0 and 1 for each feature.

In Fig. 6 the results for an over-time clustering made by DBSCAN with $minPts = 4$ and $\epsilon = 0.11$ and an outlier threshold of $\tau = 0.5$ are shown. The time series 16, 37, 48 are generated with higher deviation and 49 and 50 completely random. It can be seen that all these time series were found by our algorithm as outliers (grey). Since the data points of these time series often are outliers as well as change their cluster members, this is a correct result. However, the first two time points are assumed to be normal for time series 16. This is desired too, as it moves stable with its cluster members at this time.

Although the data points of 42, 45, 46 and 47 split from their cluster members in time point 4, they are not identified as outliers. Since they migrate together and even merge back to their former cluster members in time point 5, their behavior is not conspicuous. The sequence 42 is identified as anomalous between time points 1 and 2 (turquoise cluster), since all its cluster members migrated completely stable from time point 1 to 2.

In total, the following outlier sequences can be read from Fig. 6: $T_{3,8,16}$, $T_{1,2,42}$, $T_{1,7,37}$, $T_{1,8,48}$, $T_{1,8,49}$, $T_{1,8,50}$. All are justified and correspond to the desired result. There is one striking observation, though: Although 37 is conspicuous over the entire period, it is only found as outlier between time 1 and 7. The reason for this is that the marginal case mentioned in Sect. 4 has occurred. Since the data point of the time series was classified as noise at the very last point in time, but not at the time before, the sequence is not found by our algorithm.

6   Conclusion and Future Work

In this work we presented a robust outlier detection algorithm for multiple multivariate time series. By analyzing the cluster transitions of time series over time, we are able to identify anomalous sequences. Instead of using sliding windows, our method performs an analysis of all possible subsequences. The shown results are sound and enable a new field of research. However, there are still some interesting aspects which may be examined in future work. The most important issue is the determination of the outlier detection parameter τ. We assume an interdependence of τ and hyperparameters that are used for the clustering algorithm. Further not all intuitive outlier sequences have to be conspicuous in regard to the time series database. Considering the deviation of time series can lead to an enhanced analysis of those. Finally, it could be useful to identify whole outlier clusters. Therefore a cluster score could be computed and evaluated.

References

1. Ahmad, S., Lavin, A., Purdy, S., Agha, Z.: Unsupervised real-time anomaly detection for streaming data. Neurocomputing **262**, 134–147 (2017)
2. Ahmar, A.S., et al.: Modeling data containing outliers using ARIMA additive outlier (ARIMA-AO). J. Phys: Conf. Ser. **954**, 012010 (2018)
3. Banerjee, A., Ghosh, J.: Clickstream clustering using weighted longest common subsequences. In: Proceedings of the Web Mining Workshop at the 1st SIAM Conference on Data Mining, pp. 33–40 (2001)
4. Ben-David, S., Von Luxburg, U.: Relating clustering stability to properties of cluster boundaries. In: 21st Annual Conference on Learning Theory (COLT 2008), pp. 379–390 (2008)
5. Cheng, H., Tan, P.N., Potter, C., Klooster, S.: Detection and characterization of anomalies in multivariate time series. In: Proceedings of the 2009 SIAM International Conference on Data Mining, pp. 413–424 (2009)

6. Cho, H., Fryzlewicz, P.: Multiple change-point detection for high-dimensional time series via sparsified binary segmentation (2014)
7. Cleveland, R.B., Cleveland, W.S., McRae, J.E., Terpenning, I.: STL: a seasonal-trend decomposition procedure based on loess (with discussion). J. Official Stat. **6**, 3–73 (1990)
8. ASA Statistics Computing and Graphics: Airline on-time performance. http://stat-computing.org/dataexpo/2009/the-data.html. Accessed 15 July 2019
9. Ester, M., Kriegel, H.P., Sander, J., Xu, X.: A density-based algorithm for discovering clusters a density-based algorithm for discovering clusters in large spatial databases with noise. In: Proceedings of the Second International Conference on Knowledge Discovery and Data Mining, pp. 226–231 (1996)
10. Ferreira, L.N., Zhao, L.: Time series clustering via community detection in networks. Inf. Sci. **326**, 227–242 (2016)
11. Hill, D.J., Minsker, B.S.: Anomaly detection in streaming environmental sensor data: a data-driven modeling approach. Environ. Model Softw. **25**(9), 1014–1022 (2010)
12. Huang, X., Ye, Y., Xiong, L., Lau, R.Y., Jiang, N., Wang, S.: Time series k-means: a new k-means type smooth subspace clustering for time series data. Inf. Sci. **367–368**, 1–13 (2016)
13. Keogh, E., Lonardi, S., Chiu, B.Y.C.: Finding surprising patterns in a time series database in linear time and space. In: Proceedings of the Eighth ACM SIGKDD International Conference on Knowledge Discovery and Data Mining, KDD 2002, pp. 550–556 (2002)
14. Landauer, M., Wurzenberger, M., Skopik, F., Settanni, G., Filzmoser, P.: Time series analysis: unsupervised anomaly detection beyond outlier detection. In: Su, C., Kikuchi, H. (eds.) ISPEC 2018. LNCS, vol. 11125, pp. 19–36. Springer, Cham (2018). https://doi.org/10.1007/978-3-319-99807-7_2
15. Lin, J., Keogh, E., Fu, A., Van Herle, H.: Approximations to magic: finding unusual medical time series. In: 18th IEEE Symposium on Computer-Based Medical Systems (CBMS 2005), pp. 329–334 (2005)
16. Liu, S., Yamada, M., Collier, N., Sugiyama, M.: Change-point detection in time-series data by relative density-ratio estimation. Neural Netw. **43**, 72–83 (2013)
17. Malhotra, P., Vig, L., Shroff, G.M., Agarwal, P.: Long short term memory networks for anomaly detection in time series. In: ESANN (2015)
18. Munir, M., Siddiqui, S.A., Chattha, M.A., Dengel, A., Ahmed, S.: FuseAD: unsupervised anomaly detection in streaming sensors data by fusing statistical and deep learning models. Sensors **19**(11), 2451 (2019)
19. Paparrizos, J., Gravano, L.: k-shape: efficient and accurate clustering of time series. In: Proceedings of the 2015 ACM SIGMOD International Conference on Management of Data, pp. 1855–1870 (2015)
20. Salvador, S., Chan, P.: Toward accurate dynamic time warping in linear time and space. Intell. Data Anal. **11**(5), 561–580 (2007)
21. Sun, P., Chawla, S., Arunasalam, B.: Mining for outliers in sequential databases. In: ICDM, pp. 94–106 (2006)
22. Truong, C.D., Anh, D.T.: A novel clustering-based method for time series motif discovery under time warping measure. Int. J. Data Sci. Anal. **4**(2), 113–126 (2017)
23. Zhou, Y., Zou, H., Arghandeh, R., Gu, W., Spanos, C.J.: Non-parametric outliers detection in multiple time series a case study: power grid data analysis. In: AAAI (2018)

Applying Softmax Classifiers to Open Set

Darren Webb[(✉)]

Cyber and Electronic Warfare Division, Defence Science and Technology Group,
Edinburgh, Australia
darren.webb@dst.defence.gov.au

Abstract. Many artificial neural network classifiers assume a closed-world, where the composition of classes is fixed and known, however problems where this assumption does not hold are found in nearly every case where multi-class classification is applied. For example, in network traffic classification the actual number of classes often dramatically exceeds the number of classes known or labelled at training time. Various treatments have been proposed to adapt closed-set classifiers for application in open-set scenarios, both formal and informal. To demonstrate the effectiveness of these treatments, we conducted an empirical study using a simple example based on the MNIST digit classification problem. Our results show the various treatments make trade-offs between classification recall and precision. We demonstrate that *softmax* output equalisation during training can significantly improve performance in open set classification.

Keywords: Open-set recognition · Deep learning · Neural network

1 Introduction

Artificial neural networks have demonstrated excellent results in multi-class classification problems from many domains. The recognition of handwritten digits is widely explored and used to compare the performance of neural network approaches and architectures. The classifier learns general concepts from a database of labelled handwritten digits, and its performance is measured according to its recognition of digits in previously unseen images [6].

Many artificial neural network classifiers use a *softmax* final layer which yields a probability distribution across a fixed set of classes known at training time. This assumes a closed-world where the exact number and composition of classes is known. However, problems for which the closed-world assumption does not hold are found in nearly every case where multi-class classification is applied [10]. In the open-world the actual number of classes may dramatically exceed the number of classes known at training time. When the closed-world assumption is broken, instances from previously unknown classes must be fit to known classes, with good recall of known classes but inevitable misclassification of samples from unknown classes. This means that closed-world results look better than they really are [10].

T. D. Le et al. (Eds.): AusDM 2019, CCIS 1127, pp. 104–115, 2019.
https://doi.org/10.1007/978-981-15-1699-3_9

Our interest in this problem is motivated by our experiences applying classi-fiers to network traffic from real networks [7]. It is possible to devise closed-set traffic classification schemes (e.g. TCP, UDP, or mail, P2P, web), but in mod-ern networks these classification schemes are behaviourally orthogonal and their frequencies heavily skewed making for trivial recognition tasks. Further, in most real networks these classification schemes are still open-set. The interests of network engineers and security practitioners requires more sophisticated classifi-cation schemes where there are likely many more classes than known at training, and samples from known and unknown classes are likely to share features. There are unfortunately very few examples [1,12] where this problem is recognised and to some extent addressed. This has motivated us to experiment with different approaches to the problem.

Various treatments, formal and informal, are proposed to address the limi-tations of *softmax* to the open-set problem. Formal techniques, including Open-Max [3]), Open Classifier Networks (OCN) [9] and Evidential Deep Learning (EDL) [13], address the open-set problem by quantifying misclassification risk (or prediction uncertainty). Informal techniques, such as confidence thresholds [1,12] or inclusion of an "unknown" class, do not quantify misclassification risk, rather apply heuristic bases for rejection (such as applying a threshold to *softmax* out-puts). Consequently they may address some aspects of the open-set problem but are still vulnerable to uncertain (e.g. adversarial) inputs.

Motivated by the need to apply our network classifiers in open-set scenarios, and understand the performance implications of doing so, we report the results of our investigation of approaches to open-set recognition. In summary, this paper makes the following contributions:

- Survey and comparison of informal and formal approaches applied to an open-set scenario; and
- A new approach combining counter-example training (similar to [4]) with OpenMax to produce near-optimal classification performance in our open-set scenario.

The paper is structured as follows. Section 2 frames the open-set classifica-tion problem, describing the openly available data set and code used in our experiments. Section 3 compares oft used informal treatments with the formal OpenMax treatment. Section 4 demonstrates our contribution to improving the precision of these treatments using counter-example training. We summarise our findings in Sect. 5.

2   Breaking the Closed-World Assumption

Handwritten digit recognition is a common problem used to compare the per-formance of neural network approaches and architectures. A common data set for comparing the accuracy of approaches is the MNIST handwritten digit data set [6]. The data set consists of 60000 training images and 10000 test images of ten handwritten digits, zero to nine, with an image size of 28×28 pixels. This data set is a closed-set, containing exactly 10 known classes. However, an open-set problem can be posed by training with a subset of these classes and testing with all classes. Like our network traffic classification problems, the digits will likely share significant features across known and unknown classes.

The experiments described in this paper are based on an existing MNIST digit classifier [14], an artificial neural network consisting 3 convolutional and max-pooling layers, a fully-connected layer and *softmax* output layer. The network is trained with 1000 batches of 100 samples.

Figure 1 shows its prediction performance when trained to samples of all 10 classes, achieving overall test accuracy of 99.03%. t-SNE [8] is used to visualise the penultimate-layer responses, reducing the dimension of the responses while revealing its structure. In this case the t-SNE plot shows the penultimate-layer responses to samples of the 10 classes cluster with good separation.

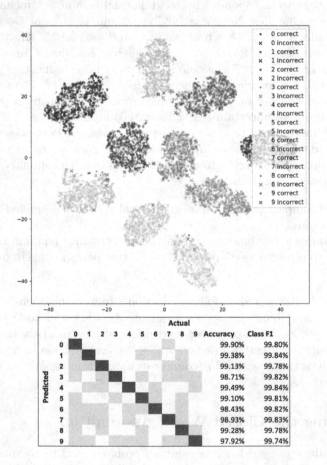

Fig. 1. Results for the initial classifier showing t-SNE plot of the penultimate layer responses and overall per-class accuracy.

To measure the effect of breaking the closed-world assumption, we train the classifier using only instances of a subset of "known" classes, and test against instances of all classes. We arbitrarily choose to train only three classes (0, 1, 2), with the remaining classes (3–9) regarded as unknown. As there are only three classes to model, the new classifier is reduced in size to minimise potential for over-fitting, with only two convolutional and max-pooling layers and a smaller fully-connected layer.

Figure 2 shows a training accuracy of 100%, however there is no mechanism to separate or reject samples of unknown classes so predictably the test accuracy against all 10 classes is only 31.40%. The t-SNE plot shows that while the known classes cluster and are evidently separable, the penultimate-layer responses to samples of the unknown classes are not. This is typical whenever a closed-world classifier is naively applied to the open-world, and consequently the problem has motivated the development of open-set treatments.

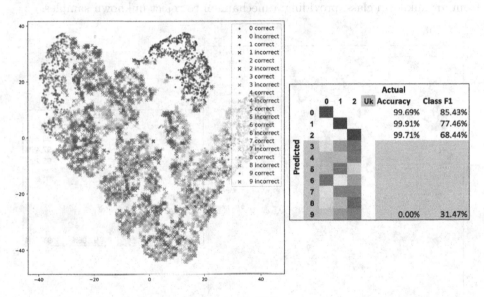

Fig. 2. Results of testing the original classifier in an open-set scenario showing the t-SNE plot of penultimate layer response, and confusion matrix with dark grey region showing the "unknown" classes not exposed during training.

3 Treatments for Open-World Classification

Various treatments are proposed to enable *softmax* classifiers to operate within open-set scenarios. The most common informal approaches are using confidence thresholds and introducing an unknown class, while OpenMax is the widely used formal approach. We now compare these treatments.

3.1   Confidence Thresholds

The simplest and most common treatment to the open-set problem is to treat the components of the *softmax* output as a confidence or uncertainty measure to which a threshold can be applied (e.g. [1,12]). If the largest component of the *softmax* probability distribution is greater than the threshold it is predicted to be of that class, else the prediction has low confidence and inferred to be of an unknown class. This reduces the boundaries around the known classes, effectively leaving an unknown space for distant examples to be regarded unknown. This is a popular technique despite [5] showing that *softmax* classifier probabilities are not directly useful as confidence estimates.

To demonstrate we apply a threshold to the maximum *softmax* component of the naive classifier in Sect. 2. The confidence threshold is informed from the distribution of confidence values after training, in this case 0.5 is chosen. A sample with a maximum *softmax* value below this threshold is considered to be from an unknown class, providing a mechanism to reject unknown samples.

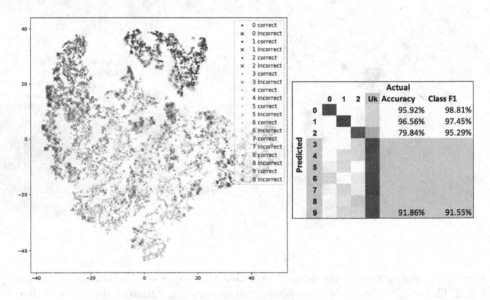

Fig. 3. Results of applying a threshold to *softmax* outputs.

Figure 3 shows the classifier achieves an overall accuracy of 91.55%. As the threshold is applied only during testing the t-SNE plot looks essentially the same as for the naive approach, however the threshold provides the means to reject samples effectively establishing decision boundaries around the samples of known classes. As there is no clear separation between the classes these decision boundaries will be imprecise. This appears particularly the case where classes share common features (e.g. in this scenario "0" and "6").

3.2   Including an "Unknown" Class

Should unlabelled training instances be available for some subset of unknown classes, there is potential to improve classifier precision. [15] includes an additional "unknown" class to capture all instances that do not match the set of known classes. Intuitively, this forces the model to more precisely define the decision boundary around known classes through the need to identify the boundary for an additional class.

To implement this approach we include an additional class in the *softmax* final layer, and train the model with the inclusion of these unlabelled instances to yield the unknown class. To reflect a likely application scenario, with classes that share features, we split the training instances into three: instances for known classes (0–2), instances for available unknown (unlabelled) classes (7–9), and instances for unavailable unknown classes (3–6). We train the classifier using only the instances for known and available unknown classes, and test using instances from all classes. The expectation is that the classifier will learn robust general concepts for the known classes, and enough about how they differ from available unknown class instances to correctly classify all unknown class instances.

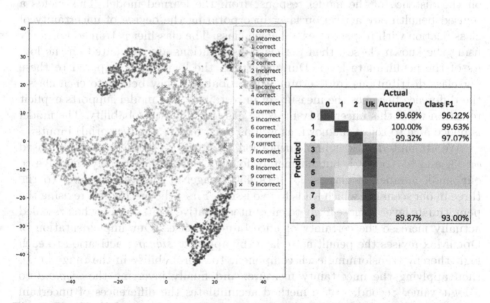

Fig. 4. Results of testing the classifier with an additional "unknown" class. The confusion matrix shows the "unknown" samples used to training the "unknown" class in light grey.

Figure 4 shows the classifier in this case achieves an accuracy of 92.96% while classification of the unknown class is 89%. The t-SNE of penultimate layer responses shows the known samples cluster and are separated from the available unknown samples, but misclassification occurs mainly in the region between. These responses are established during training and hence the unavailable unknown samples could produce similar responses to any of the available classes.

The central problem is best summed up by [16], that all positive examples are alike but each negative example is negative in its own way. The model learns to respond strongly to features of the available unknown samples but these do not generalise to all unknowns, and so the unavailable unknowns may draw strong responses from features of any class.

We have shown these informal approaches can achieve some degree of recognition in the open set. However, the techniques described do not satisfy the formal definition of open set recognition offered in [11], as they focus on model performance and ignore misclassification risk. We next consider a formal approach that enables rejection based on prediction uncertainty.

3.3 OpenMax Classification

OpenMax [3] is an approach to open recognition that models both model performance and misclassification risk. It employs a decaying probability model based on the distance of the model response from the learned model. This yields a revised penultimate activation layer incorporating the degree of uncertainty of classification with respect to each known class. The classifier is trained normally using the known classes, then per-class distributions are calculated on the logits of the penultimate layer. During testing, the logits are compared to these per-class distributions to determine the probability they belong to each class - the difference becomes the measure of uncertainty. The model supports explicit rejection when this uncertainty measure has the highest probability. The model is hence able to reject unknown as well as uncertain (e.g. adversarial) inputs.

We attempted various implementations and hyper-parameter settings based on [3] and its code [2], but ultimately found OpenMax extremely sensitive to our small number of classes. [3] uses 450 ImageNet classes compared to the three in our scenario, which leads to two issues. First, the method for revising the penultimate layer is more likely to encounter negative logits, which when rescaled actually increase the certainty of uncertain responses. Our implementation of OpenMax revises the penultimate layer by applying *sigmoid* activation to each logit thereby transforming each component to a probability in the range [0, 1], then applying the uncertainty measure, and finally inverting the *sigmoid* to a logit value. Second, as the method accumulates the differences of uncertain responses to determine an overall uncertainty value, the sum of small logits may never yield an uncertainty above the other logits. In the case the logits of the penultimate layer are all below zero, we deem the outcome uncertain and returns a default uncertainty logit above zero.

The results in Fig. 5 show the OpenMax-based classifier achieves an overall accuracy of only 72.21%, well below the other approaches tested. The t-SNE plot shows the revised logits do not appear to separate known and unavailable unknown classes. We believe this is likely due to significant overlap in features between these classes. The model has not learned what differentiates

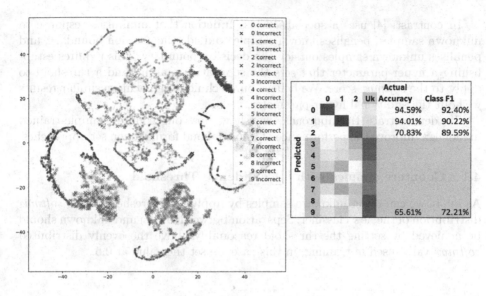

Fig. 5. Results of testing OpenMax on a model trained with samples of the three known classes.

these classes, only the features that differentiate the known classes, hence the penultimate layer has similar responses to samples from known and unknown classes.

4   A Better Open-World Classifier

The open-set recognition techniques described provide high recall but poor precision. To improve separation between samples of known and unknown samples, and hence assist precision, we developed an approach similar to [4] using the available unknowns as counterexamples (also referred to as background examples [4] or negative examples [11]). Intuitively, we train the classifier to yield a strong response to samples from known classes, and weak responses to counterexamples. These weaker responses are easier to reject, hence the model is more robust to misclassification of unknown instances.

Counterexample training is achieved by manipulating the probability distribution during training. During training, the optimiser seeks a solution that yields a *softmax* distribution close to the desired probability distribution. Typically this probability distribution has the maximum probability of one for the correct class and minimum probability of zero for the incorrect class. In our approach, we evenly distribute the probability across the known classes (i.e. the probability must sum to one so the three known classes each have probability 0.33). During testing, the trained model yields a peaked probability distribution for samples of known classes, and a flat distribution for samples of unknown classes.

In contrast, [4] uses a specialised loss function that minimises response to unknown samples, penalises known samples outside the decision boundary, and penalises unknown samples outside the decision boundary. This requires establishing a hyper-parameter that sets the separation margin, and a threshold to apply to the *softmax* score. We believe our technique can achieve similar results without the additional hyper-parameter.

To demonstrate the approach, we show how our counterexample-trained model can improve the performance of informal and formal open set approaches.

4.1 Counterexamples with a Confidence Threshold

As in [4], we can reject unknown samples by applying a threshold to the *softmax* distribution of the new classifier. Separation between known and unknown should be achieved by setting the threshold reasonably above the evenly distributed *softmax* value used in training, in this case we set the value at 0.5.

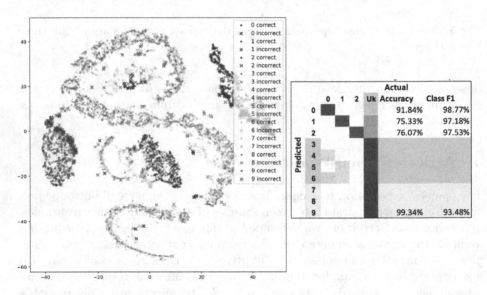

Fig. 6. Results of testing the confidence-threshold on counterexample-trained model. The confusion matrix shows the classes used for counter-example training in light grey.

The results in Fig. 6 show the classifier achieves accuracy of 93.48%, a 2% improvement over applying a threshold without counterexamples. Compared to the previously studied treatments, this approach performs better for instances of previously-unseen classes with clear separation between the responses of known and unknown classes. However, the chosen threshold preferences classification of unknown samples at the cost of rejecting samples from known classes.

4.2   Counterexamples with OpenMax

The fundamental expectation of OpenMax is different inputs produce different responses. In our scenario with a small number of trained known classes and overlapping unknown classes, the penultimate layer responses are relatively close. The results in Sect. 3.3 demonstrate that OpenMax cannot discriminate known from unknown under these conditions, however the more precise responses achieved by training with counterexamples can assist. Figure 7 shows OpenMax classification with a counterexample-trained model achieved 97.68% accuracy with lowest per-class accuracy of 90.89%. This is a substantial improvement over applying a confidence threshold, with the added benefit of not needing to establish additional hyper-parameters.

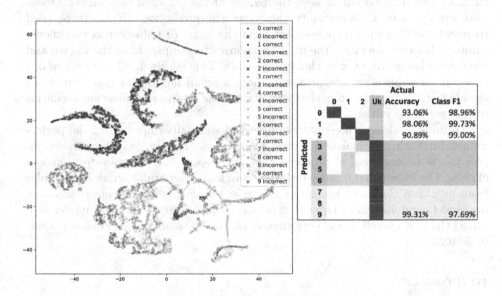

Fig. 7. Results of testing OpenMax with counterexample-trained model.

The t-SNE plot for the revised penultimate layer logits shows the known classes are clustered and clearly separated from unknown samples. In contrast to the earlier OpenMax model, the counterexample-trained model achieves better separation between the responses of known and unknown classes that OpenMax is better able to exploit.

5   Conclusions

Many real-world classification problems are open-set in nature, however *softmax* classifiers are designed and trained under a closed-world assumption. The resulting models have no mechanism to reject samples from unknown classes.

Network traffic classification is inherently open-set, motivating our interest in this problem.

We investigated empirically a number of treatments that enable the application of *softmax* classifiers to open-set scenarios. Common treatments such as thresholds and the addition of an "unknown" class can yield acceptable recall but typically with poor precision as the learned model has not learned separable features. We tested each treatment using an openly available data set [6] and code [14], and used t-SNE on the penultimate-layer responses to visualise and explain the effect of each treatment. We have applied the techniques to network traffic classifiers with comparable results.

Formal approaches to open-set quantify misclassification risk, while informal approaches do not. While a formal approach is highly desirable, our implementation of OpenMax did not achieve the results of the informal treatments. OpenMax incorporates an uncertainty value into the prediction which can be used to reject samples with responses not from the same distribution as the known samples, however we found the model responses to samples from the known and unknown classes were not so clearly separable. This is principally a result of our constrained application scenario, with only a small number of classes to learn defining characteristics, but characteristic of the traffic classification scenarios in real-world networks.

Training models with counterexamples substantially improved model performance for both informal and formal approaches. Training with counterexamples yields a classifier that produces stronger responses to samples from known classes. The known and unknown classes were highly separable, enabling samples from unknown classes to be more easily rejected. For our constrained scenario, our novel combination of OpenMax on the counterexample-trained model produced the best overall model performance of 97.68% against the closed-set result of 99.03%.

References

1. Aceto, G., Ciuonzo, D., Montieri, A., Pescapè, A.: Mobile encrypted traffic classification using deep learning. In: 2018 Network Traffic Measurement and Analysis Conference (TMA), pp. 1–8 (2018)
2. Bendale, A.: abhijitbendale/OSDN (2016). https://github.com/abhijitbendale/OSDN. Accessed 17 June 2019
3. Bendale, A., Boult, T.E.: Towards open set deep networks. In: 2016 IEEE Conference on Computer Vision and Pattern Recognition (CVPR), pp. 1563–1572 (2016)
4. Dhamija, A.R., Günther, M., Boult, T.E.: Reducing network agnostophobia. CoRR abs/1811.04110 (2018). http://arxiv.org/abs/1811.04110
5. Hendrycks, D., Gimpel, K.: A baseline for detecting misclassified and out-of-distribution examples in neural networks (2016)
6. LeCun, Y., Cortes, C.: MNIST handwritten digit database (2010). http://yann.lecun.com/exdb/mnist/
7. Li, Y., et al.: Deep content: unveiling video streaming content from encrypted wifi traffic. In: 2018 IEEE 17th International Symposium on Network Computing and Applications (NCA), pp. 1–8 (2018)

8. van der Maaten, L., Hinton, G.: Visualizing data using t-SNE. J. Mach. Learn. Res. **9**, 2579–2605 (2008). http://www.jmlr.org/papers/v9/vandermaaten08a.html
9. Prakhya, S., Venkataram, V., Kalita, J.: Open set text classification using convolutional neural networks. In: International Conference on Natural Language Processing, December 2017. http://par.nsf.gov/biblio/10059464
10. Rocha, A., Goldenstein, S.: Multi-class from binary - divide to conquer. In: VISAPP (2009)
11. Scheirer, W., Rocha, A., Sapkota, A., Boult, T.: Toward open set recognition. IEEE Trans. Pattern Anal. Mach. Intell. **35**(7), 1757–1772 (2013). https://doi.org/10.1109/TPAMI.2012.256
12. Schuster, R., Shmatikov, V., Tromer, E.: Beauty and the burst: remote identification of encrypted video streams. In: USENIX Security Symposium (2017)
13. Sensoy, M., Kaplan, L., Kandemir, M.: Evidential deep learning to quantify classification uncertainty. In: Proceedings of the 32Nd International Conference on Neural Information Processing Systems, NIPS 2018, pp. 3183–3193. Curran Associates Inc., USA (2018). http://dl.acm.org/citation.cfm?id=3327144.3327239
14. Sopyła, K.: ksopyla/tensorflow-mnist-convnets (2017). Accessed 1 May 2019
15. Zhang, J., Chen, X., Xiang, Y., Zhou, W., Wu, J.: Robust network traffic classification. IEEE/ACM Trans. Netw. **23**, 1 (2014). https://doi.org/10.1109/TNET.2014.2320577. http://nsclab.org/nsclab/esi/tonzhang2015.pdf
16. Zhou, X.S., Huang, T.S.: Small sample learning during multimedia retrieval using biasmap. In: CVPR (2001)

An Efficient Risk Data Learning with LSTM RNN

Ka Yee Wong and Raymond K. Wong[(✉)]

School of Computer Science and Engineering,
University of New South Wales, Sydney, Australia
kayee.wong@unsw.edu.au, wong@cse.unsw.edu.au

Abstract. The use of big risk data for risk management has evolved into a concern within financial services industry in recent years. Whether the quality of big risk data can be relied upon is to be ascertained till 2019. To facilitate the measurement and prediction of data quality, we propose an efficient approach to slide a piece of data from the big risk data and a model to train divergent Long Short-Term Memory ("LSTM") Recurrent Neural Networks ("RNNs") with various algorithms. The network is evaluated by the improvement in network run time, prediction accuracy and relevant error. This enables financial institutions to identify potential data risks instantly for earlier mitigation soon.

Keywords: Long Short-Term Memory · Recurrent Neural Network · Big data

1 Introduction

In Feb 2019, the government in Australia demanded for the restoration of trust in financial system after several mis-conduct of the banks [1]. This is in view of recent scandals: In April 2018, the Prudential Regulator refused a bank's corporate risk data due to data inaccuracy & incompleteness [2]. In Feb the same year, the Royal Commission challenged banks for mis-conduct arising from the poor quality of risk data [3].

In recent years, banks in Australia put much effort on the application of AI technology to reduce bank compliance costs, called RegTech ("Regulatory Technology"), by utilizing the power of big data [4]. This is also a trend for them to apply machine learning to big data for future management [5]. Both AI or machine learning are trusted to be used to swiftly filter the bad quality from the big data and slide the good quality for use. For example, used to predict the data quality with machine learning algorithms.

How to justify a data as good or bad is referenced to the data quality dimensions set out by the regulator, ARPA – CPG 235 [6]. The most onerous task is that the big data is stored in multiple repositories. To analyse and predict them takes tremendous amount of time [7]. They have deviating formats and the size are immense such as market risk ("MR"), credit risk ("CR"), operational risk ("OR") and liquidity risk ("LR").

In order to use the big risk data for risk management, we propose a machine learning model to learn with every bit of the information to improve the predictive power. The model is implemented by a deep neural network, LSTM RNN, identifying complex patterns in large data sets to make accurate prediction. It is tested by scenarios to ascertain the effectiveness in terms of the saved run time, enhanced accuracy and

© Springer Nature Singapore Pte Ltd. 2019
T. D. Le et al. (Eds.): AusDM 2019, CCIS 1127, pp. 116–128, 2019.
https://doi.org/10.1007/978-981-15-1699-3_10

improved prediction error – through the deployment of another algorithm and network. The saved run time is the time saved in training the neural network for the prediction.

The prediction of risk data quality requires the processing of previous and future cases for the reduction of likelihood of a future occurrence. Big risk data accumulated over years in banks. A massive network is required. To fit this, we select LSTM RNNs to explore information from the past & future [8] to build an enormous network and model long-term temporal dependencies [9]. To apply memory between batches let it remember memory across long sequences to obtain control over the network state. Thus, this is used to forecast the quality to identify data risk earlier for mitigation [10].

2 LSTM-RNN Learning Approach

Pursuant to the CPG 235, the data quality ("DQ") is to be measured by several dimensions: (a) accuracy; (b) completeness; (c) consistency; (d) timeliness; (e) availability; and (f) fitness for use [6]. These dimensions are mapped to the real-world data quality issues as defined in Table 1. The issues are implanted into the dataset of this paper.

Table 1. Data quality issues mapped to the CPG 235 data quality dimensions

Dimensions	Real world 10 common DQ issues
Accuracy	Translation issue may be an omission of the translation for an amount (from US$ to AU$)
	Transformation issue may be an omission of the transformation of date format (from dd/mm/yy to mm-dd-yyyy)
Completeness	Incomplete values include abnormal postal codes inconsistent with the standard digits for a country
Consistency	Data-mismatch should be investigated attributable to inaccurate record in one or more databases
	Duplicated fields are to be de-duplicated to avoid extra inconsistent record including duplicated identifier for the same customer in the database
Timelines	Stale records need to be updated to keep up-to-date and any obsolete records over a period be removed
Availability	Missing value is either due to an omission or a value to be imputed later. Both need correction
Fitness-for-use	Redundant data is unnecessary and be removed. An example is a potential client previously rejected by the bank, but his/her data is still kept in the database
	Unreasonable data occurs if a customer hit with the anti-money laundering is still held in the database
	Invalid data include some transactions made with the client before an account is opened for him/her

2.1 The Dataset

Real-world data quality issues were regularly announced by risk experts (e.g. market risk [11, 12], credit risk [13], liquidity risk [14] and operational risk [15, 16]. In this paper, we: (a) analyze the structure of these and summarize the key characteristics; (b) make-reference to other methodologies related to the organization of data defects [17]; and (c) magnify the common data quality issues in the dataset.

We synthesize a dataset for use in the model: http://ndb.cse.unsw.edu.au/regtech/datasets/201904. It contains 1 million banking customer instances that capture all possible non-compliance scenarios according to the CPG 235. It has 132 data features (called "data elements") belonging to 4 risk databases including MR, CR, OR and LR. Each database contains 33 features in which 8 are static data and 25 are dynamic data. They include corporate and individual data and the values are discrete. Some data features with quality issues are extracted to the following (with examples).

- Market risk – Asset Amount (251527, na, ' ', 838); Nationality (' ', tbc, JPY, GBP, AUD); MR Segments (Retail Bank, Private Bank, Wholesale Bank); Customer Risk Rating (H, M, L, OnBoarding, P, Q);
- Credit risk – Collateral Amount (29397, 6727, tbc, ' '); TimeStamp (3/10/2015 6:09, 6/02/2008 3:15, 16/06/2018 4:34); Guarantor ID Number (123272, 32416, tbc); Product Price (725, 85, 3089, na);
- Operational risk – Event Date (06.25.2009, 10.05.2012, 1.29.2018); Residual Legal Liability (1385, 12, 3307, 715); Control Factors (system control, na, others, regular review); Loss Multiplier (1, ' ', 1.4, 2); and
- Liquidity risk – Settlement Date (4/11/2014, 13/12/2009, 19/08/2018); NAV (tbc, 871942, 17914, ' '); Liquidity Rate (0.1039, 0.5103, 0.9975); Number of Trades (1685, 13, na, ' ').

Data features with data quality issues are assigned with a quality score before classifying into a quality rank, as an input into the network for the data quality prediction.

2.2 Data Labeling

Prior to inputting data into the networks, we labeled them as depicted in Fig. 1.

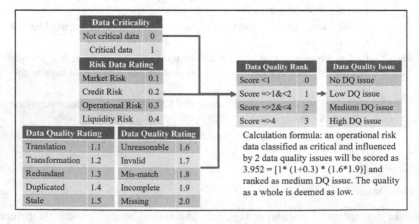

Fig. 1. Data labeling

Firstly we: (a) label data as 1 or 0 to indicate if it is critical; (b) assign a risk data rating (0.1 to 0.4) & a data quality rating (1.1 to 2) based on the types of risk data & quality issues respectively; (c) classify data quality scores (<1, =>1 & <2, =>2 & <4 and =>4) into 4 ranks (no/ low/ medium/ high data quality issue) and compare the ranks with the prediction made in the experiment. Data criticality, all ratings & rankings are generally assigned by risk experts in the industry.

2.3   Data Scoring Approach and Data Pre-processing

The score is a multiplication of the rating for data quality issues: Data Criticality Factors * (1 + Risk Data Rating) * Data Quality Rating. Input data is pre-processed. The abnormal data deviating from the standard (such as missing values) is identified and its quality score is normalized to a value from 0 to 1 by a min-max scaler. The predicted quality level exceeding the threshold computed by the machine learning algorithm is classified into a bad quality whereas that within the threshold is categorized into a good quality. The threshold is a portion of the length of input sequence. The output value of 1 represents a quality issue exceeding threshold, and value 0 infers that there is no issue.

3   LSTM-RNN Learning Model

In the model, we train the LSTM RNN with our dataset. The data is split into two – 70% for training the network and 30% for network evaluation. In network setup, we define the number of previous time steps (look_back) as inputs to predict the data quality for next time while time steps are ticks of time. Default time is set at 1 and next time is set at (t + 1). The input array: [samples, time steps, features] is imported into the network. Also, we frame the quality issue as a one-time step for each sample. In total, there is a layer with 1 input, a hidden layer with 4 LSTM blocks and an output layer of 1. The activation function of the network is sigmoid, and the batch size is 1.

For another network, the LSTM RNN with memory between batches, which is used to compare the initial network, we reset the state in the network after each batch.

3.1   Data Profiling

The big risk data is stored in large-scale repositories [5]. Retrieving takes time. In view of this, we conduct a data profiling to select data clusters for testing in model:

In market risk database, the data element of "Asset Amount" is selected.

- Asset Amount (in number) is classified into 4 categories (1, 2, 3 & 4 corresponding to the amount of <85,000, between 85,000 & 385,000, over 385,000 & all amounts). Out of these categories, the number of records is around 10%, 30%, 60% and 100% of the database.

Besides, the data element of "Nationality" is selected.

- Nationality (in several options) is classified into 4 categories: 1, 2, 3 & 4 corresponding to a group of countries (CAD, SGD & EUR), another group of countries (AUD, CAD, CNY, CZK, EUR, GBP, HKD, JPY, MYR, NZD & SGD), a group of countries (excluding category 1 & 2) & all countries. Out of these, the number of records is 9%, 34%, 65% and 100% of the database.

In credit risk database, "CollateralAmt" is selected.

- CollateralAmt (in amount) is classified into 4 categories (1, 2, 3 & 4 corresponding to the amount of <85,000, between 85,000 & 385,000, over 385,000 & all amounts). Out of these categories, the number of records is 10%, 30%, 60% and 100% of the database.

In operational risk database, "EventDate" is chosen.

- EventDate (in date format of mm.dd.yyyy) is classified into 4 categories (1, 2, 3 & 4 corresponding to the dates later than 19 Feb 2017, the dates later than 19 Feb 2015, the dates later than 19 Feb 2012 & all dates). Out of these categories, the number of records is 16%, 36%, 66% and 100% of the database.

In liquidity risk database, "SettlementDate" is chosen.

- SettlementDate (in date format of DD/MM/YYYY) is classified into 4 categories (1, 2, 3 & 4 corresponding to aging period: between 1 & 2 years, 1 & 4 years, 1 & 7 years & all years). Out of these categories, the number of records is 10%, 30%, 60% and 100% of the database.

Above 4 risk databases are consolidated into the integrated dataset. Following the profiling, the least percentage of the data from the entire database (either 9%, 10% or 16%) can be used as a base for the comparison of the network run time in experiments.

3.2 Network and Relevant Algorithm

There are memory blocks connected via layers in LSTM RNN. The LSTM block has components to make a memory for recent sequences and contains gates to manage block's state and output, and operates on input sequence, as shown in Fig. 2 [18, 19].

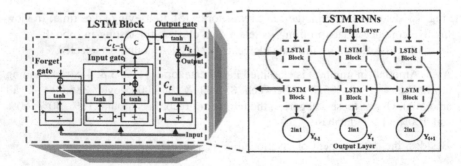

Fig. 2. Network architecture

In LSTM, each gate uses sigmoid activation units to control if they are triggered or not, for changing the state and adding information flow. The LSTM is combined with the RNN by predicting the quality based on previous predictions and the information learned, unlike conventional feed-forwarding network where inputs are independent. This is the sequence handling of inputs in the RNN.

RNNs use information from past time steps and from future ones to extend networks to the sequential data, as computed in [18, 20]: $h_t = \phi(W_h h_{t-1} + W_x x_t)$ where a sequence of hidden states $(h_1, \ldots, h_T)$ is output at time step (t) from the input sequence of vectors $(x_1, \ldots, x_T)$ where W_h is recurrent weight matrix, W_x is input-to-hidden weight matrix and ϕ is an activation function.

LSTM is a structure of RNNs incorporating gates (input, forget & output) to manage block's state & output and operating on an input sequence as computed in equation below when i_t, f_t and o_t are input, forget & output gates:

$$i_t = \text{sigmoid}(W_{xi}x_t + W_{hi}h_{t-1}) \tag{1}$$

$$f_t = \text{sigmoid}(W_{hf}x_t + W_{hf}h_{t-1}) \tag{2}$$

$$c_t = f_t \odot c_{t-1} + i_t \odot \tanh(W_{xc}x_t + W_{hc}h_{t-1}) \tag{3}$$

$$o_t = \text{sigmoid}(W_{hx}x_t + W_{ho}h_{t-1} + W_{co}c_t) \tag{4}$$

$$h_t = o_t \odot \tanh(c_t) \tag{5}$$

where c_t is the cell, $(\cdot)$ is logistic sigmoid function and tanh is the hyperbolic tangent function. By integrating memory between batches (BN), the network's hidden becomes:

$$h_t = \phi(BN(W_h h_{t-1} + W_x x_t)) \tag{6}$$

The optimization algorithm is ADAGRAD [21], a variant of stochastic gradient descent:

$$\theta_{t+1,i} = \theta_{t,i} - \frac{\eta}{\sqrt{G_{t+1,i} + \epsilon}} \bullet g_{t,i\bullet} \tag{7}$$

where η is learning rate at each time step t for every parameter θ_i based on the past gradients computed for θ_i and $g_{t,i}$ is the partial derivative of the objective function. $G_t \in \mathbb{R}^{d \times d}$ is a diagonal matrix where each element i is a sum of squares of gradients up to the timestep whereas ϵ is a smoothing term avoiding division by 0 (on order of $1e - 8$). Hence, ADAGRAD uses a different learning rate for every parameter θ_i at each time step. It sets the value as 0.01 and adapts the rate to parameters, performing low rates for parameters with frequently occurring features and high rates for those with infrequent features. Then, we vectorize the implementation by performing a matrix vector product $\odot$ between G_t and g_t (denoting gradient at t):

$$\theta_{t+1} = \theta_t - \frac{\eta}{\sqrt{G_t} + \varepsilon} \odot g_{t\bullet}$$ (8)

The ADAGRAD converges in batch setting like this [22]:

Algorithm

Input: Tolerance $\varepsilon > 0$. Initiate $x_0 \varepsilon$, $\mathbb{R}^d b_0 > 0$, $j \leftarrow 0$

repeat

$\quad j \leftarrow j + 1$

$\quad b_j^2 \leftarrow b_{j-1}^2 + \| \nabla F(x_{j-1}) \|^2$

$\quad x_j \leftarrow x_{j-1} - \frac{1}{b_j} \nabla F(x_{j-1})$

until $\| \nabla F(x_j) \|^2 \le \varepsilon$

---

4 Experiment

Python v3.5 is used with Keras library & tensorflow as backend to train the networks on a system with the processor of i.7-7500U CPU@2.9 GHz, OS of 64-bit and Win 10.

4.1 Results

Utilizing the algorithm of ADAGRAD, we select data elements from MR to train the network with the dataset of 10%, 30%, 60% & 100% of the entire database. Elements are asset amount and nationality. Testing results are shown in Figs. 3 and 4.

(a) Cumulative Runtime for Diff. % of Database with ADAGRAD

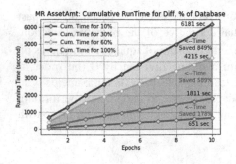
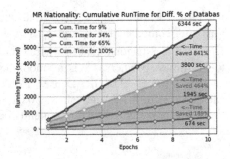

(b) Runtime Range with ADAGRAD

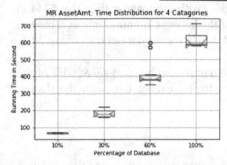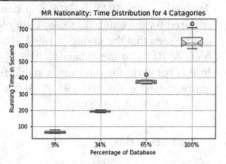

(c) Runtime Trend with ADAGRAD

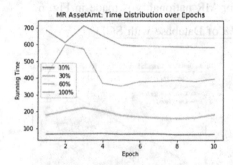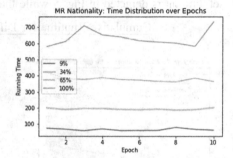

Fig. 3. Runtime for MR asset amount   **Fig. 4.** Runtime for MR nationality

- Referring to the MR asset amount in Fig. 3, much run time is saved in the network with a dataset which is 10%, 30%, 60% of the entire database. Using 10% of the network running time as a base, we find that the percentage of time saving is 178%, 589% & 849% for the dataset having 30%, 60% & 100% of the data from entire database. It is attributable to the total runtime of 1811, 4215 & 6181 s ("sec") respectively. By selecting a small portion of data for prediction (by 10%), we reduce a massive amount of network run time. It simply requires 651 s.

- Inside the network, the run time on average is stable over 10 epochs for the dataset with 10% & 30% of the entire database (over 65 s and 181 s) except that with 60% & 100%. The run time for the dataset with 60% of data from entire database rises to a high level (598 s) at 2nd epochs but returns to a normal level till the end of epoch (393 s) whereas that for the dataset having 100% of the entire database is unstable – dropping at 2nd epochs (611 s) and bouncing back to a high level at 3rd epochs (711 s) before decreasing at an average level (600 s) at 4th epochs.

- The average runtime is visualized in Fig. 3. They are for the dataset sourcing 10%, 30%, 60% & 100% of the data from entire database. The widest range of run time lie in 100% dataset sourcing all data from entire database.

- Shifting to the nationality in Fig. 4, we achieve a similar result. Significant amount of runtime is saved with the dataset of 9% data sourcing from entire database. The runtime is limited to 674 s. It is also saved for the dataset having 34% & 65% of the database – only 1945 & 3800 s in comparison with the lengthy time (by 6344 s) required for the entire database. Using 9% of the network running time as a base, it can be seen that the saving is 189%, 464% & 841% for the dataset containing 34%, 65% & 100% of the database.

- In this network, the run time is stable for the datasets with 9%, 34% & 65% of the entire database (by 67, 195 & 380 s respectively) except that with the dataset having 100% of the data.

4.2 Test Scenarios

To check if the network performance can be improved, we conduct additional tests.

Scenario 1 - We train the network with another algorithm, SGD. The run time for MR asset amount is depicted in Fig. 5 while that for MR nationality is made in Fig. 6.

Cumulative Runtime for Diff. % of Database with SGD

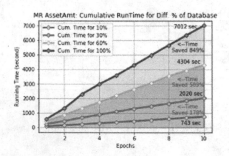 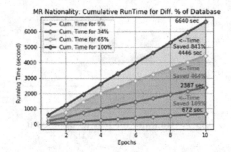

Fig. 5. Runtime for MR asset amount   **Fig. 6.** Runtime for MR nationality

Again, we notice that much time is saved by applying the SGD. The run time for the dataset with 10%, 30%, 60% of the entire database is 743, 2020 & 4304 s when compared with the total time of 7012 s. Given this, the percentage of extra run time is 178%, 589% & 849% for the dataset of 30%, 60% & 100% of the data from entire database respectively assuming the runtime of dataset with 10% of the entire database is used as a base. Additionally, this situation is the same as that of the nationality. The variance is the percentage of extra time – 189%, 464% & 841% for the dataset sourcing 34%, 65% & 100% of the data from entire database.

Scenario 2 – We compare the LSTM RNN with another network, LSTM RNN with memory between batches. To test it, we select asset amount from MR to apply both algorithms (ADAGRAD & SGD) to the new network, as given in Figs. 7 and 8.

Cumulative Runtime for Diff. % of Database with Another Network

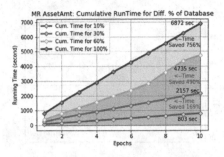 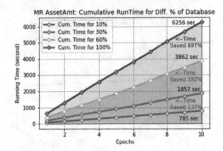

Fig. 7. Runtime for MR asset amount   **Fig. 8.** Runtime for MR asset amount

To train the LSTM RNN with memory batches with ADAGRAD, the saved time is 169%, 490% & 756% for the dataset having 30%, 60% & 100% of the entire database. When compared with the initial network, the saving is similar. By applying SGD, the saving is explicit - 137%, 392% & 697%. Then, we understand the prediction, accuracy and loss of the new network. Selecting a dataset with 10% of data for testing, the accuracy is found high (69.40%) and loss is found low (0.616) with respect to the network with ADAGRAD in Fig. 9. For the new network with SGD, the accuracy is highly low (0.89%) while the loss is high (11.187) in Fig. 10. As a result, we prefer the ADAGRAD.

(a) Prediction of LSTM RNN with Memory

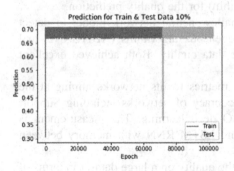 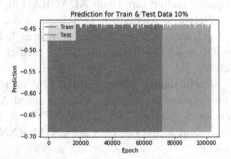

(b) Accuracy & Validated Accuracy

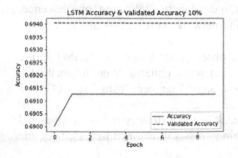 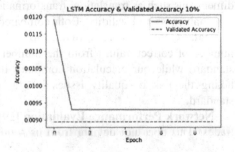

(c) Loss & Validated Loss

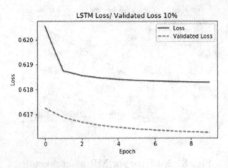
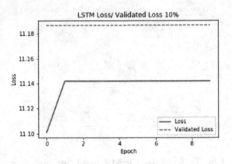

Fig. 9. Network performance (ADAGRAD)   **Fig. 10.** Network performance (SGD)

5 Related Work

In the prediction of big risk data quality, [23] studied how machine learning enhanced the network performance in terms of the predictive/classification accuracy. Our experiments demonstrate how to drive the performance with a deep learning network.

Network Architecture [24] applied a deep architecture model using auto-encoders to represent traffic flow features for prediction. Our model is a LSTM RNN modelling long-term temporal dependencies and remembering memory across long sequences. It successfully discovered latent traffic flow feature representation but ours find out a superior architecture (with ADAGRAD algorithm) for the quality prediction.

Network Algorithm [25] innovated a shallow neural network model to detect colon cancer but our model compares heterogeneous algorithms (ADAGRAD and SGD) in LSTM RNNs to rapidly predict the data quality. Both achieved excellent results.

Performance Measurement [26] utilized metrics for its networks aiming at the consumption prediction – training speed & accuracy of networks including Support Vector Regression (SVR), local SVR & H2O deep learning. The measurement is equivalent to us, but ours are LSTM RNN and LSTM RNN with memory between batches.

Data Quality Dimensions [27] measured the quality on a large dataset in terms of the accuracy, completeness and consistency. Ours include these as well as other dimensions such as translation, transformation, redundancy, duplication, obsolescence, reasonableness and validity. Both compared the network with various algorithms.

Data Quality Score Calculation [28] calculated the accuracy by dividing the number of correct values from the number of observations based on the ISO 25012 standard while our calculation computes the data score under a scientific method – taking the risk of quality issues into account after alignment with the CPG 235 standard.

Network Performance Evaluation [29] back-tested a strategy to assess simulated trades. This does not deviate from us – utilizing cross-validations to check back the accuracy and loss of network prediction. Both demonstrated the success of processing a cluster of the big data for a prediction within a reasonable time of few hours.

6 Conclusion

We demonstrate an improvement in the performance of machine learning networks for the big risk data prediction – run time, accuracy and loss. This enables financial institutions to find the good and bad quality of data swiftly from the big data. Besides, they can rely on the model to analyse the quality issues with a fraction of data. Given this, they can remediate data earlier for risk management. Additionally, the initial problem stated in this paper can be remedied. To accomplish this, we propose a systematic approach to slide the big risk data for quality prediction and a model to forecast the quality in LSTM RNNs automatically. In future, we will test other big data in machine learning to help resolve other real-world issues.

References

1. Frydenberg, H.J.: Restoring trust in Australia's financial system. Australian Government, The Treasury, pp. 3–42 (2019)
2. Frost, J.: APRA rejected CBA home loan data as inaccurate and incomplete. Financial Review: Business, Banking and Finance, p. 1 (2018)
3. Yeates, C.: Banks dive as UBS raises home loan concerns. Sydney Morning Herald: Banking & Finance, pp. 1–2 (2018)
4. Eyers, J.: CBA want to use AI to tackle fraud and cyber attacks. Business Insider, pp. 1–2 (2016)
5. Härle, P., Havas, A., Kremer, A., Rona, D., Samandari, H.: The future of bank risk management. Mckinsey & Company, pp. 1–32 (2015)
6. APRA: Prudential Practice Guide CPG 235 – Managing Data Risk. Australian Prudential Regulation Authority (APRA), pp. 1–13 (2013)
7. Upadhyay, A.: How long does it take to complete a big data or data science project in real time on large data sets?. Quora, p. 1 (2017)
8. Zhou, P., et al.: Attention-based bidirectional long short-term memory networks for relation classification. In: Proceedings of the 54th Annual Meeting of the Association for Computational Linguistics, pp. 207–212 (2016)
9. Zhu, W.T., et al.: Co-occurrence feature learning for skeleton based action recognition using regularized deep LSTM networks. In: Proceedings of the Thirtieth AAAI Conference on Artificial Intelligence, Association for the Advancement of Artificial Intelligence, pp. 3697–3702 (2016)
10. Mikolov, T., Joulin, A., Chopra, S., Mathieu, M., Ranzato, M.A.: Learning longer memory in recurrent neural networks. In: International Conference on Learning Representation, arXiv preprint arXiv:1412.7753 [cs] (2014)
11. KPMG: Equity market risk premium – research summary. KPMG, pp. 3–7 (2018)
12. CFI: Market risk premium. Corporate Finance Institute (CFI), pp. 2–5 (2018)
13. Moody's Analytics: Credit risk calculator. Moody's, pp. 3–7 (2018)
14. KPMG: Basel 4: the way ahead, operational risk, the new standardized approach. KPMG, pp. 3–9 (2018)

15. Migueis, M.: Is Operational risk regulation forward-looking and sensitive to current risks?. Board of Governors of the Federal Reserve System, pp. 1–7 (2018)

16. IOSCO: Recommendations for liquidity risk management for collective investment schemes. The Board of the International Organization of Securities Commissions (IOSCO), pp. 1–20 (2018)

17. Josko, J.M.B., Oikawa, M.K., Ferreira, J.E.: A formal taxonomy to improve data defect description. In: Gao, H., Kim, J., Sakurai, Y. (eds.) DASFAA 2016. LNCS, vol. 9645, pp. 307–320. Springer, Cham (2016). https://doi.org/10.1007/978-3-319-32055-7_25

18. Ergen, T., Kozat, S.S.: Efficient online learning algorithms based on LSTM neural networks. IEEE Trans. Neural Netw. Learn. Syst. **29**(8), 3772–3783 (2017)

19. Palangi, H., et al.: Deep sentence embedding using long short-term memory networks: analysis and application to information retrieval. IEEE/ ACM Trans. Audio Speech Lang. Process. **24**(4), 694–707 (2016)

20. Fan, Y.C., Qian, Y., Xie, F.L., Soong, F.K.: TTS synthesis with bidirectional LSTM based recurrent neural networks. In: Conference of the International Speech Communication Association, pp. 1964–1968 (2014)

21. Kingma, D.P., Ba, J.L.: ADAM: a method for stochastic optimization. In: International Conference on Learning Representations. arXiv preprint arXiv:1412.6980, [cs.LG] (2014)

22. Ward, R., Wu, X.X., Bottou, L.: AdaGrad stepsizes: sharp convergence over nonconvex landscapes. In: International Conference on Machine Learning, pp. 6677–6686 (2019)

23. AI-Jarrah, O.Y., Yoo, P.D., Muhaidat, S., Karagiannidis, G.K., Taha, K.: Efficient machine learning for big data: a review. Big Data Res. **2**(3), 87–93 (2015)

24. Lv, Y.S., Duan, Y.J., Kang, W.W., Li, Z.X., Wang, F.Y.: Traffic flow prediction with big data: a deep learning approach. IEEE Trans. Intell. Transp. Syst. **16**(2), 865–873 (2015)

25. Chen, H.M., Zhao, H., Shen, J., Zhou, R., Zhou, Q.G.: Supervised machine learning model for high dimensional gene data in colon cancer detection. In: IEEE International Congress on Big Data, pp. 134–141 (2015)

26. Grolinger, K., Capretz, M.A.M., Seewald, L.: Energy consumption prediction with big data: balancing prediction accuracy and computational resources. In: IEEE International Congress on Big Data, pp. 157–164 (2016)

27. Serhani, M.A., E1 Kassabi, H.T., Taleb, I., Nujum, A.: An hybrid approach to quality evaluation across big data value chain. In: IEEE International Congress on Big Data, pp. 418–425 (2016)

28. Taleb, I., E1 Kassabi, H.T., Serhani, M.A., Dssouli, R., Bouhaddioui, C.: Big data quality: a quality dimensions evaluation. In: International Conferences on Ubiquitous Intelligence & Computing, Advanced and Trusted Computing, Scalable Computing and Communications, Cloud and Big Data Computing, Internet of People and Smart City Innovation, pp. 759–765. IEEE (2016)

29. Ruta, D.: Automated trading with machine learning on big data. In: IEEE International Congress on Big Data, pp. 824–830 (2014)

Using Transfer Learning to Detect Phishing in Countries with a Small Population

Wernsen Wong[✉], Yun Sing Koh[✉], and Gillian Dobbie

School of Computer Science, The University of Auckland, Auckland, New Zealand
wwon129@aucklanduni.ac.nz, {ykoh,gill}@cs.auckland.ac.nz

Abstract. An increasing number of people are using social media services and with it comes a more attractive outlet for phishing attacks. Phishers curate tweets that lead users to websites that download malware. This is a major issue as phishers can gain access to the user's digital identity and perform malicious acts. Phishing attacks also have a potential to be similar in different regions, perhaps at different time periods. We investigate the use of transfer learning to detect phishing models learned in one region to detect phishing in other regions. We use a semi-supervised algorithm to train a model on a US based dataset that we then apply to New Zealand. First, we evaluate how effectively transfer learning can be used in different regions to detect potential phishing attacks on online social networks in real time. Secondly, we investigate the different phishing attacks and discuss the differences in phishing attack features detected for different countries. We have collected a real world Twitter dataset over 6 months and show that we are able to detect phishing successfully using US phishing models despite only a low level of phishing occurring in smaller populations such as New Zealand.

Keywords: Phishing detection · Transfer learning · Model transfer

1 Introduction

Due to the rise of social media as a mainstream platform for communication, attackers have shifted their focus to these social media. A key example in September 2014, the Internet in New Zealand had a disastrous meltdown due to the spread of malware downloading spam, which lured users to click on links which claimed to contain Hollywood star photos, but in actual fact directed users to download malware to perform DDoS attacks. Social media statistics report about 500 million tweets are created daily on Twitter alone and 326 million monthly active users [13]. Subsequently, online social media and security companies are combating spammers to make online social network platforms phishing-free. For example, Google uses a blacklisting service called Google Safe Browsing service. Twitter introduced BotMaker, which uses blacklist filtering as a component in their detection system. However, blacklists fail to protect victims from new spam

© Springer Nature Singapore Pte Ltd. 2019
T. D. Le et al. (Eds.): AusDM 2019, CCIS 1127, pp. 129–140, 2019.
https://doi.org/10.1007/978-981-15-1699-3_11

due to the time delay between when new phishing spam and when it is registered on the blacklist. Research shows that more than 90% of victims may visit a new spam link before it is blocked by blacklists. Nowadays phishers are sophisticated and adapt to game the system with fast evolving content and network patterns to avoid being detected. It is challenging for existing anti-phishing systems to quickly respond to newly emerging patterns for effective phishing detection.

To build a phishing detection model, a sufficient amount of data is needed to build an accurate and non-biased model. Building a general phishing detection model gives us a general perspective of the type of phishing links produced at a particular time. Insights to understand the problem of spam phishing attacks on social networks from a geographic regional point of view, would be beneficial for the understanding of how the phishing attacks are designed specifically for different regions. In order to look at this problem, we first need to analyse and collect information from countries in different geographic regions.

In small countries it is difficult to collect enough phishing data to build models when phishing models are adapting so quickly. Thus we need to use models found in larger countries to capture any new phishing models and transfer the learning from another domain to allow detection of phishing in smaller populations. We proposed a complete model transfer and an instance based transfer techniques. We plan to use "appropriate" knowledge from a different source domain, for a target domain. In our specific case we are looking at leveraging the phishing detection model from a large population, such as the United State's model (US model), to create a prediction model for a smaller populated country, such as New Zealand (NZ model). Our system is a two stage system. The first stage involves identifying what instances of the training data are transferred to aid in training the smaller population's model. We used our technique called adaptive phishing detection technique (ADPT) to build a model to detect phishing. We show that the performance of ADPT is reliable on large populations in real time. This approach has a higher F1, recall, and precision value than most existing research. Some current research uses features that do not exist in real time such as number of retweets. We showed that ADPT is a suitable base classifier for our problem. We do not concentrate on this portion of the research as the focus of this paper is geared towards the effectiveness of transfer learning to model phishing in a smaller population. The second stage involves training our adaptive phishing detection technique (ADPT) on the newly adjusted smaller population's data to more accurately identify phishing attacks. We investigated two simple transfer learning mechanisms, using (i) a pre-trained model from a larger dataset (US data) on the smaller dataset (NZ model) and (ii) an instance transfer learning technique which transfers related instances.

The main contributions of this paper are as follows. We propose two transfer learning techniques to use phishing models from a larger population countries to detect phishing in smaller population countries. The first technique is to use a pre-trained model from the United States (US) data on the New Zealand (NZ) data. The second technique uses instance transfer learning from the US data to help build the NZ model.

The paper is organized in the following manner. In Sect. 2, we set up our related work. In Sect. 3, we present our data collection and feature extraction mechanism. In Sect. 4 we discuss our phishing detection technique and Sect. 5 discuss the proposed transfer learning. Section 6 discusses the experiments and results. Section 7 contains the concluding remarks.

2   Related Work

This section explores both the phishing detection research and transfer learning research areas.

Phishing Detection. Recently there have been various research in the area of phishing detection. The broad area of research includes concentrating on feature selection and engineering used in the models, advancement of the modeling techniques, and sentiment analysis. In feature selection and engineering, they concentrate on how different features can improve the classification of spammers [4]. In the advancement of the modeling techniques, researchers attempt to use different classifiers such as Support Vector Machines, Decision Trees and Naive Bayes to detect spammers [1,10].

Transfer Learning. Transfer learning is normally used when it was too expensive to rebuild new models based on the new target domain, or when it is impossible to collect the training data needed and rebuild the models. In such cases, knowledge transfer or transfer learning between task domains would be desirable. There has been various research that has focused on transfer learning [12]. Overall transfer learning can be divided into inductive transfer [2,5], transductive transfer [3] and unsupervised transfer [6].

There are different applications of transfer learning such as sign language recognition [8] and text classification [14]. Instance-transfer is an approach that directly applies training data from another dataset to training the target model. An example of this is TrAdaBoost [5], this technique aims to transfer one source of data to help in training the target model.

3   Data Collection and Feature Extraction

Tweet Collection. Twitter has several ways of retrieving tweets as they are posted, one is through the streaming API. We use the Streaming API as it allows tweets to be collected in real-time. This also allows us to filter out tweets that are not of interest. In this research, we are only interested in tweets that are in English, contain a redirect link outside of Twitter and have occurred in New Zealand, Australia, Singapore or The United States for our experiments. During a six month period we have crawled over 1,478,639 tweets, and of those 1,002,799 are retweets and 475,840 non re-tweets.

Tweet Labelling Policy. To determine if a tweet was phishing or non-phishing, a policy is applied to generate the true labels. This is needed as we need to have a suitable ground truth to validate our classification results. After 14 days if the tweet is deleted, we label the tweet as phishing. The 14 days time period provides reasonable time for Twitter to remove malicious tweets. We have manually reviewed a sample of the removed tweets and the majority lead to sites that contain malicious ads and malicious download links.

Feature Extraction. Previous studies on phishing detection on Twitter show that features based on the URLs, the tweeter and the tweet itself can be used to determine if a tweet is malicious or not. These features have been studied in previous literature [4,10] however we wish to detect tweets in real-time and some of the features do not exist in a real-time context. These include favourites count and re-tweets, the risk of excluding these features is a loss of information however we choose to use an algorithm to make up for this loss.

Twitter User Features. We can create features from the tweet creator (tweeter) that can be used to detect phishing tweets along with other features. We use 7 identified Twitter user's features including *Follower count, Following count, Account age, Lists count, Favourites count, User description exists,* and *User is verified.* For example, phishers may create multiple new Twitter accounts that are used to tweet out the malicious links. Thus we may see that a low *Account age* and low *Follower count* would be an indicator tweets posted by this tweeter are phishing.

Twitter Tweet Features. Phishing Tweet features change over time as the phishers get more creative in evading the existing phishing detection systems. In Twitter, hashtags and mentions are important as they determine the visibility and search ranking of tweet. By mentioning genuine users, phisher can improve their tweets' visible to users who follow those genuine users which is exactly what phishers wants. Additionally, there is extra metadata for a tweet that can only be retrieved via the Twitter API, which can not be seen from the surface of Twitter such as if the *Geolocation exists* which also be useful. We identified eight tweet features including *Tweet length, Favourite count, URL count, Hashtag count, Mention count,* and *Geolocation exists.*

Twitter URL Features. There have been case studies that have revealed that URL features contribute to the identification of phishing sites. Many phishing sites abuse browser redirection to bypass blacklists by chaining together multiple redirections, therefore the number of redirections between the specified URL and the final URL is another feature we collected. The identified URL features including *URL length, URL dot count, Domain length,* and *Redirection count.*

4   Adaptive Phishing Detection Technique

We have proposed a technique, Adaptive Phishing Detection Technique (ADPT), to classify tweets in real-time by using different classifiers. By combining classifiers together, ADPT leverages various individual techniques to diversify the types of phishing detected by the individual classifiers. ADPT uses Stacked Generalization to combine several classifiers of different types into an ensemble by passing their results into an overarching classifier. ADPT's architecture is seen

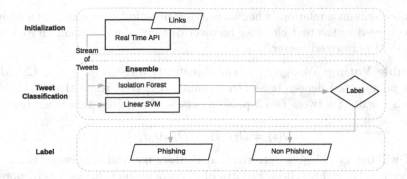

Fig. 1. APDT architecture

in Fig. 1. Our technique is a modified version of a stacked generalization instead of using the same type of training data. ADPT splits the training data set based on feature type. A tweet consists of different features, for example the tweets are *free text* unstructured data type, and contain characteristics of the *tweets and URL features* which are mostly numeric and structured data. ADPT uses the classifier that allows us to best model the data.

The *Initialization* step collects the data from Twitter's Streaming API, extracts features and performs the model training; the *Tweet Classification* uses an ensemble to classify the tweet with a phishing or non-phishing label. ADPT leverages both Isolation Forest (IF) [11] for tweet analysis and Support Vector Machines (SVM) for the free text analysis [4] in the ensemble. Our phishing technique is preferred over other methods as it is designed to work in real-time, as an online technique. As a result, we are only focusing on tweet features that can be collected and analysed in real time.

Isolation Forest Outlier Detection: IF is an outlier detection technique that uses random forest to detect outliers. It attempts to isolate observations by randomly selecting a feature and then selecting a random split value between the minimum and maximum value of the feature selected. This is done recursively until an outlier threshold is reached. The outliers are decided by the number of splits, where trees that produce shorter path lengths than other samples are more likely to be outliers [11]. For example, if we looked at a single tweet's feature set then it could start on the tweet's text length and split a tree based on a random value. The majority of the tweets would be in one partition while the minority or potential phishing would be in the other partition.

LSVM Classifier: SVM is a classifier where the goal is to find a hyper-plane that optimally separates the training data into two portions [4]. We use the linear kernel as the transformed tweet text, which is linearly separable. Linear SVM (LSVM) have been used before for sentiment analysis and spam detection [16]. LSVM requires the tweet's text to be transformed into text features. We use the bag of words (BOW) [9] model, where the tweet is considered a 'Bag' containing words. We apply a uni-gram model and a bi-gram model, which allows us to keep more semantic relations when identifying phishing. Naïve Bayes could have also been used as the text classifier however our initial experiments have shown that LSVM performed better.

Ensemble Voting: We aggregate the classification result of IF and LSVM for a data point i and denote it as $D_{IF}(i)$ and $D_{LSVM}(i)$ respectively. Based on the assignment of a tweet i to a phishing class, the aggregation is expressed as follows:

$$D(i) = D_{IF}(i) \text{ or } D_{LSVM}(i)$$

LSVM will detect phishing patterns that IF does not and vice versa. For example, recently there have been reports of cryptocurrency hacks where popular compromised accounts are positing phishing tweets asking for cryptocurrency. IF would classify these as phishing due to the behaviour being normal. However the text is similar to other phishing text that have been posted before, allowing SVM to classify this tweet as phishing. It is comparable to benchmark techniques such as SVM [1,4] despite the loss in real-time features such as tweet favourites and retweets count. There is a reduced amount of phishing compared to the number of phishing attacks captured by previous research. In May 2018, Twitter's systems identified and challenged more than 9.9 million potential spam or automated accounts per week. This is up from 6.4 million in December 2017, and 3.2 million in September 2017 [15].

The main motivation behind using ADPT comes from this technique performing better than the other benchmark techniques such as SVM [4] and PDT [10]. We used prequential evaluation to evaluate our model, where instances are first used to test, and then to train. Both SVM and PDT do not perform well in a real time stream setting as they were designed with features that were not real time, such as the number of times a tweet has been retweeted. The problem of small and scarce data is not easily resolved with any traditional phishing detection technique or outlier detection technique. Thus we further investigated the possibility of using simple transfer learning mechanism to boost the effectiveness of learning models for a smaller population, such as NZ.

5   Transfer Learning for Phishing Detection

In this section we formally define the problem of transfer learning. We proposed two approaches of transfer learning to address the phishing detection problem.

Problem Statement. In this section we introduce the problem of transfer learning following the definition by [5]. Given a domain D is made of feature space $\mathcal{X}$ and a marginal distribution $P(X)$, where $\mathcal{X} = x_1, \ldots, x_n$, and $x_i \in \mathcal{X}$. A task $\mathcal{T}$ made up of a label space $\mathcal{Y}$, and a function $f : \mathcal{X} \to \mathcal{Y}$. The learning task $\mathcal{T}$ for the domain $\mathcal{D}$, amounts to estimating a classifier function $f : \mathcal{X} \to \mathcal{Y}$, from a given training data $D = (x_1, y_1), \ldots, (x_n, y_n) | x_i \in \mathcal{X}, y_i \in \mathcal{Y}$ that best approximates f, based on a certain criteria. We denote $\mathcal{D}_T = (\mathcal{X}, P_T(X))$ as target domain for which we can learn target class $\mathcal{T}_T = (\mathcal{Y}, f_T)$, from the target training data $D_T = (x_1^T, y_1^T), \ldots, (x_n^T, y_n^T)$. We denote $\mathcal{D}_S = (\mathcal{X}, P_S(X))$ as source domain for which we can learn source class $\mathcal{T}_S = (\mathcal{Y}, f_S)$, from the source training data $D_S = (x_1^S, y_1^S), \ldots, (x_n^S, y_n^S)$. We aim to improve the learning of the target classifier by using the knowledge from the source task $\mathcal{T}_S$ in the source domain $\mathcal{D}_S$. This is also known as inductive transfer learning. The source domain is the US dataset, and the target domain is the New Zealand dataset.

Performing transfer learning may be advantageous over traditional machine learning in the situation when the size of the training data D_T is very small, and also relative to the size of the training data D_S. In traditional machine learning this would normally suffer from overfitting problems as the small dataset would not be able to capture the true distribution. We leverage the idea of inductive instance transfer learning, which attempts to standardise the learning problem by transferring the knowledge from a source domain that has a large training dataset for learning the source task [5]. We use this concept to reduce the overfitting of the NZ D_T set, as it is very small in terms of phishing by transferring select data from the US dataset D_S.

Pre-trained Model Using US Dataset. In this technique, we pre-trained a model using the source data and transfer the model, f_S that is trained on the source dataset D_S and apply it to the test set of the target domain D_T, essentially $\mathcal{T}_T = (\mathcal{Y}, f_S)$. We trained the model using the ADPT approach. Despite this technique being a naive solution, it can solve the insufficient data problem. An assumption is that the source domain distribution is similar to the target domain which is sometimes the case. For example, in the case of phishing, similar phishing scams do originate from US and eventually appear in NZ. We believe that using a model that is trained on the US dataset will be able to detect similar types of phishing in the NZ dataset.

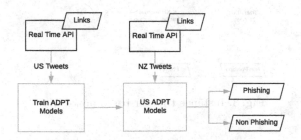

Fig. 2. Pre-trained US model transfer

The proposed architecture can be seen in Fig. 2. We will first train the ADPT technique on the US Twitter data and save the model. We then apply this pre-trained model on the NZ Twitter data to then classify each of the tweets as either phishing or non-phishing. Our method does not require fine-tuning the components. It is a conceptually simple and effective transfer model.

Instance Transfer Learning. In the previous section we used the complete pre-trained model on the source domain, US phishing dataset, and assume that the distribution of the target domain, NZ phishing dataset is the same. We can optimize this approach by only training on data that is relevant in the target domain which will improve performance of the target domain model.

Our second approach is based on instance transfer learning. We transfer data from the source domain D_S to train the model of the task target f_T. The selection of the data of the source domain is based of an intuitive selective bootstrapping approach, whereby we are only considering data points that are similar to the target data. Ideally this would reduce negative transfer from the source domain. In our case we select the source domain data to transfer by finding the most similar phishing tweets that are closest to the task domain phishing tweets. We use k-Nearest Neighbour (k-NN) [7] as shown in Fig. 3, with Euclidean distance as the distance measure to achieve this. The initial settings are $k = 10$. For each instance in the target domain (x^T) we select k nearest points from source domain D^S for modeling. To prevent outliers in the source dataset from being selected due to the fact that the nearest neighbours of an instance in the target data do not have common neighbours to the source dataset, we introduce an ϵ boundary threshold. We will only consider an instance, x^T that is a k nearest neighbour if $d(x^T, x^S) < \epsilon$, whereby x^S is the instance in the target data, and $d(.)$ is the distance function. If we have an instance in the NZ dataset that had no close neighbours in the US dataset then this would prevent us transferring unrelated instances that may reduce the accuracy of the target model.

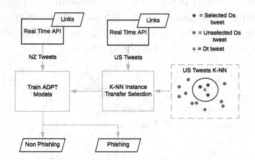

Fig. 3. Instance transfer based on k-NN

6   Experiments

We discuss the characteristics of the datasets we collected. We then discuss the performance of the transferred models. Code and further information can be found here[1].

---

[1] https://github.com/wernse/ADPT-Instance-Transfer.

Datasets. In order to evaluate the application of the transfer learning methods we have collected a US and NZ based dataset over 6 months from 2018-09-01 to 2019-01-29. A total of 498 tweets were collected for NZ and 190,084 for the US. The breakdown is shown in Table 1.

The NZ dataset is small compared to the US dataset, as it is only 0.26% of the US dataset. In terms of the amount of phishing as a percentage of the total tweets that occurred in the region, NZ has 18.07% while the US has 34.04%. US has a higher rate of phishing that occurs compared to NZ, which may be due to a smaller population so they are less attractive to attackers.

Table 1. Summary of the phishing datasets

Datasets	Start date	End date	New Zealand non-phishing	New Zealand phishing	United States non-phishing	United States phishing
VS 1	2018-09-01	2018-11-01	162	49	64516	34224
VS 2	2018-11-01	2018-12-05	97	26	36812	18279
VS 3	2018-12-05	2019-01-29	149	15	24050	12203
Total			408	90	125378	64706

Performance Evaluation. We compare three methods to evaluate the performance of proposed transfer learning model: a baseline method, a complete pretrained model transfer method and the inductive transfer method. The baseline method involves training on local NZ data without transfer learning, as discussed before this may lead to overfitting and we hypothesize that this will detect the least amount of phishing. The pure model transfer method will train on the US dataset and be applied to the NZ dataset, and we hypothesize that will be more accurate than the baseline method. The final method will use selected parts of the US dataset in combination with the NZ dataset to train a model that we will use to detect phishing attacks in NZ, and we hypothesize this technique will be the most effective at detecting phishing.

The results are shown in Table 2. The *Train* and *Test* columns reference the validation set (VS) defined in Table 1, and the remaining columns are defined as follows: f_{NZ+US} model is based on the instance transfer learning from the instances from the US dataset to train the NZ dataset, $f_{NZ,NZ}$ model trained on NZ and tested on NZ, $f_{US,NZ}$ model is the pre-trained model of the US tested on NZ dataset. We notice that the results are consistent with our hypotheses, where inductive transfer learning detects the most phishing in every experiment run. In the first experiment the inductive transfer learning method f_{NZ+US} had an average recall of 0.7147 while the baseline and pure model transfer had 0.5173 and 0.5642 respectively. There is a trade-off in terms of recall and precision, in the first experiment even though f_{NZ+US} has a higher recall, the precision is lower than the other two methods. When the US tweets were transferred they varied in distribution depending on the selection boundary. This can also be seen in the k-NN selection in the following sub section where the recall increases as the selection boundary gets larger but reduces the precision.

Table 2. Performance results for phishing detection

Train	Test	F1			Precision			Recall		
		f_{NZ+US}	$f_{NZ,NZ}$	$f_{US,NZ}$	f_{NZ+US}	$f_{NZ,NZ}$	$f_{US,NZ}$	f_{NZ+US}	$f_{NZ,NZ}$	$f_{US,NZ}$
VS1	VS2	0.6091	0.6429	0.5588	0.4762	0.6091	0.4524	**0.8933**	0.6923	0.7308
VS1	VS3	0.1714	0.2000	0.1739	0.1500	0.1714	0.129	**0.4000**	0.2667	0.2667
VS2	VS1	0.6034	0.6977	0.6250	0.5833	0.6034	0.5556	**0.7143**	0.6122	0.7143
VS2	VS3	0.4493	0.2581	0.1667	0.1622	0.4493	0.1212	**0.7561**	0.2667	0.2667
VS3	VS1	0.6379	0.7143	0.7216	0.5714	0.6379	0.7292	**0.7551**	0.6122	0.7143
VS3	VS2	0.4545	0.6939	0.6207	0.4762	0.4545	0.5625	**0.7692**	0.6538	0.6923
	Mean	0.4876	0.5345	0.4778	0.4032	0.4876	0.5173	**0.7147**	0.5173	0.5642
	Std	0.1751	0.2385	0.2438	0.1968	0.1751	0.1964	**0.1657**	0.1964	0.2308

Investigation of k-NN for Instance Transfer Learning. To choose the correct threshold for the instance transfer k-NN boundary we ran a sensitivity analysis, as shown in Table 3. The results match the recall and precision trade-off. When the threshold is set to 0.1 we get very similar neighbours so the precision is 0.6971 but with a recall of 0.5997 it does not capture as much phishing compared to 0.005, which has a wider selection boundary. In this case we selected 0.01 as we wish to prioritize recall as we would like to implement a phishing detection system that assists moderators by warning them of phishing rather than automatically mis-classifying tweets as phishing.

Table 3. Sensitivity analysis of k-NN threshold

ϵ Threshold	F1	Precision	Recall
0.1	0.6384	**0.6971**	0.5997
0.05	0.6236	0.6406	0.6173
0.01	0.5742	0.4876	**0.7147**
0.005	**0.5964**	0.5450	0.6716

Pairwise Attribute Relationship Analysis. To investigate the difference in phishing attributes we have display graphs that are based on pairwise attributes in Fig. 4.

It shows the linear regression curves of phishing and non phishing in the US dataset as well as the NZ phishing data points. The red dots and line are the US phishing, blue dots and line are US non-phishing and the green dots are the NZ phishing seen in Fig. 4. As you can see the betas of the pairwise attribute relationships are visually different for phishing and non-phishing. For example, phishing tends to have a linear relationship when the *domain url dot count* increases then so does the *tweet length* compared to non-phishing that

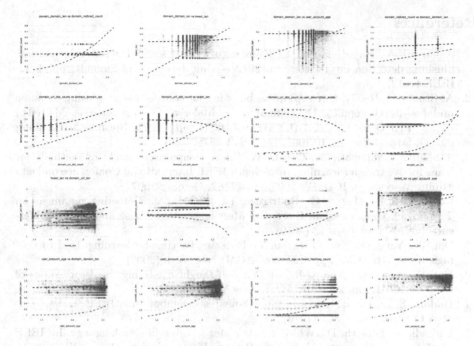

Fig. 4. Tweet pairwise attributes **Note:** US Phishing is red, US non-phishing is in blue and NZ Phishing is in green. (Color figure online)

decreases in *tweet length* as the *domain url dot count* increases. A reason this may happen is when phishers have a templated message when phishing, we found templated spam tweets such as "We're #hiring! Read about our latest #job opening here General Labor". In comparison, genuine users often do not template their messages when sending out tweets.

7   Conclusion and Future Work

We have applied our phishing detection technique to a real-world scenario and applied transfer learning techniques to enable us to detect phishing in small population countries such as NZ. We have also shown that inductive transfer learning performs better than the benchmark technique and the direct model transfer as we are able to capture a wider range of phishing tweets from the domain source that reduced the bias of the model. We have also shown interesting insights behind the difference in attributes between US phishing and US non phishing as well as the difference between US phishing and NZ phishing. In the future, we will investigate using a combination of ADPT's drift detection algorithm and inductive instance transfer learning.

Acknowledgements. This   research   is   supported   by   InternetNZ   (Grant No:IR170017).

References

1. Aggarwal, A., Rajadesingan, A., Kumaraguru, P.: PhishAri: automatic realtime phishing detection on Twitter. In: 2012 eCrime Researchers Summit, pp. 1–12. IEEE (2012)
2. Al-Stouhi, S., Reddy, C.K.: Adaptive boosting for transfer learning using dynamic updates. In: Gunopulos, D., Hofmann, T., Malerba, D., Vazirgiannis, M. (eds.) ECML PKDD 2011. LNCS (LNAI), vol. 6911, pp. 60–75. Springer, Heidelberg (2011). https://doi.org/10.1007/978-3-642-23780-5_14
3. Arnold, A., Nallapati, R., Cohen, W.W.: A comparative study of methods for transductive transfer learning. In: Seventh IEEE International Conference on Data Mining Workshops, ICDMW 2007, pp. 77–82, October 2007
4. Benevenuto, F., Magno, G., Rodrigues, T., Almeida, V.: Detecting spammers on Twitter. In: Collaboration, Electronic Messaging, Anti-Abuse and Spam Conference (CEAS), vol. 6, p. 12 (2010)
5. Dai, W., Yang, Q., Xue, G.R., Yu, Y.: Boosting for transfer learning. In: Proceedings of the 24th ICML, pp. 193–200. ACM, New York (2007)
6. Dai, W., Yang, Q., Xue, G.R., Yu, Y.: Self-taught clustering. In: Proceedings of the 25th ICML, pp. 200–207. ACM, New York (2008)
7. Dudani, S.A.: The distance-weighted k-nearest-neighbor rule. IEEE Trans. Syst. Man Cybern. **4**, 325–327 (1976)
8. Farhadi, A., Forsyth, D., White, R.: Transfer learning in sign language. In: IEEE Conference on Computer Vision and Pattern Recognition, pp. 1–8. IEEE (2007)
9. Go, A., Huang, L., Bhayani, R.: Twitter sentiment analysis. Entropy **17**, 252 (2009)
10. Jeong, S.Y., Koh, Y.S., Dobbie, G.: Phishing detection on Twitter streams. In: Cao, H., Li, J., Wang, R. (eds.) PAKDD 2016. LNCS (LNAI), vol. 9794, pp. 141–153. Springer, Cham (2016). https://doi.org/10.1007/978-3-319-42996-0_12
11. Liu, F.T., Ting, K.M., Zhou, Z.H.: Isolation forest. In: 2008 Eighth IEEE International Conference on Data Mining, pp. 413–422. IEEE (2008)
12. Pan, S.J., Yang, Q.: A survey on transfer learning. IEEE Trans. Knowl. Data Eng. **22**(10), 1345–1359 (2010)
13. Smith, K.: 58 incredible and interesting twitter stats and statistics (2019). https://www.brandwatch.com/blog/twitter-stats-and-statistics
14. Wang, P., Domeniconi, C., Hu, J.: Using wikipedia for co-clustering based cross-domain text classification. In: Eighth IEEE International Conference on Data Mining, pp. 1085–1090. IEEE (2008)
15. @yoyoel, @delbius: How Twitter is fighting spam and malicious automation (2018). https://blog.twitter.com/official/en_us/topics/company/2018/how-twitter-is-fighting-spam-and-malicious-automation.html
16. Zangerle, E., Specht, G.: Sorry, I was hacked: a classification of compromised Twitter accounts. In: Proceedings of the 29th Annual ACM Symposium on Applied Computing, pp. 587–593. ACM (2014)

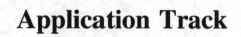

Application Track

Classifying Imbalanced Road Accident Data Using Recurring Concept Drift

Robert Anderson[✉], Yun Sing Koh, and Gillian Dobbie

School of Computer Science, University of Auckland, Auckland, New Zealand
rand079@aucklanduni.ac.nz, {ykoh,gill}@cs.auckland.ac.nz

Abstract. In New Zealand, road accident casualties have been increasing. Factor analyses and time series analyses show what types of accidents result in casualties, but the results from the analysis can become outdated. We propose a stream classification framework with drift detection to signal and adapt when the factors associated with crash casualties change over time. We propose a drift detection framework, G-mean Adaptive drift Detection (GAD), which adapts a classifier threshold to maximise G-mean. This metric rewards maximising accuracy on each class while keeping these accuracies balanced. As a result, GAD can make concept drift in the minority class easier to detect. We also propose a recurring concept classification framework, G-mean Concept Profiling Framework (GCPF), which reuses previously trained classifiers and uses GAD's approach to drift detection. Through experimentation, we show GAD improves G-mean without increasing false positive drift detection on imbalanced synthetic and real world datasets. We also show GCPF achieves better G-mean than other state-of-the-art stream classification approaches on the NZ crash data set.

1 Introduction

The International Traffic Safety Data and Analysis Group (IRTAD), a permanent working group at the OECD, state in their 2018 annual report that road deaths are reducing across member countries [12]. From 2010 to 2016, 27 of 32 countries reduced or kept road deaths stable, and 24 of 29 reporting countries had stable or reduced deaths in 2017 compared to 2016. However, they report that New Zealand has had four consecutive years of increases in road deaths from 2013, with a 15.6% increase in 2017 compared to 2016. This increase in road accidents has come to the attention of the New Zealand Government which has announced a $1.4bn programme to make New Zealand's roads safer through nationwide awareness campaigns, improvements of road infrastructure and adjusting speed limits [15]. However, the factors that result in casualties may change over time. With continual changes in road use behaviour (*e.g.*, rapid uptake of taxi applications and e-scooters) we cannot assume that factors that were important in previous years are important now unless the data shows that to be the case. An ideal solution would be to learn from crash data (such as location, speed limit, number of vehicles involved) as it arrives and predict whether

© Springer Nature Singapore Pte Ltd. 2019
T. D. Le et al. (Eds.): AusDM 2019, CCIS 1127, pp. 143–155, 2019.
https://doi.org/10.1007/978-981-15-1699-3_12

a given example of a crash would be likely to result in casualties. This solution should recognise when factors that lead to casualties have changed, signal this, and adapt to the change. Stream classification frameworks with drift detection can do this. Classifiers determine the strength of association between crash factors and casualties while drift detectors can provide information about when this relationship changes. Understanding when change occurs allows policymakers to validate that road safety campaigns are still based on current data.

Road accident data in New Zealand is imbalanced, since only a minority of crashes result in casualties: between 25%–34% annually since 2000. To use data stream classification, we must first address the wider problem of drift detection on imbalanced data streams. Explicit drift detection has been recognised as a challenging task on class imbalanced data as it generally relies on drops in classification accuracy to identify drift [5,16]. However, classifiers may have low accuracy on a minority class, so a drop in accuracy can be difficult to detect. Improving minority class accuracy can address this.

Our first contribution addresses the challenge of drift detection on imbalanced data, which is required if we are to usefully signal drift on crash data. The G-mean Adaptive drift Detection (GAD) is a drift detection framework that improves G-mean of classifiers in imbalanced binary-class datasets. G-mean is the harmonic mean of precision and recall [13], and is increased by maximising accuracy on each class while keeping these accuracies balanced. GAD improves classifier G-mean by adapting the classification threshold (*i.e.,* evidence required for our classifier to identify an instance as the minority class) when drift occurs. The threshold is set at a level that would have achieved the best G-mean on recent data prior to the drift. Due to imbalance in our dataset, using accuracy as a metric would reward classifying all but the most obviously dangerous crashes as resulting in no casualty. A metric like G-mean will reward frameworks that can successfully identify borderline cases that are more likely to result in casualties. Many recent drift detectors, *e.g.* [7], provide performance guarantees based on sample size, which may not hold with oversampling techniques used to address class imbalance, such as those in [16]. GAD performs no post-processing so performance guarantees will still hold.

New Zealand road accident data appears to have cyclical elements and it is possible that factors associated with casualties in crashes will recur over time. Recurring concepts frameworks store classifiers which they reuse after drift detection to improve accuracy [6,10]. For our second contribution we adapt an existing stream classification framework, Enhanced Concept Profiling Framework (ECPF) [2], to create the G-mean Concept Profiling Framework (GCPF). We implement GAD within ECPF to create GCPF, a recurring concepts framework that maximises G-mean over imbalanced data streams. This classification framework can detect drift and signal when it occurs. Through reusing classifiers, our framework can achieve better minority class accuracy faster, allowing drifts to be more easily detected within the minority class. Over the NZ road crash dataset, we show that GCPF achieves better G-mean than state-of-the-art recurring concepts and ensemble stream classification frameworks. We show

GCPF generalises well to a related problem, by achieving better G-mean than other approaches when classifying whether a severe/fatal casualty occurred for crashes over 7 years in road accident data from the UK.

2   Related Work

Road safety is a widespread concern, and there is a rich body of research on the subject. However, Lavrenz et al. [14] identify that due to challenges in finding appropriate datasets and a lack of understanding in the field, time-series modelling has traditionally been underused in favour of factor analysis. The authors identify that road usage problems tend to change over time. Time-series models can represent change over time, but as they generally learn from a static batch of data, the user must monitor to find when learnt models are outdated.

A data stream classifier learns from new examples as they arrive and then discards them. This allows scaling to datasets of any size. Using explicit drift detection, these approaches can detect when concept drift occurs and take remedial action such as training a new classifier. A concept is the distribution of the dataset attributes X (or crash factors), the class y (whether the crash resulted in a casualty), and how they relate to each other $Pr(y|X)$ at a given point in the stream. A concept drift occurs when the factors related to crash casualties change or the proportion of crashes result in a casualty $Pr(y)$ changes. For example, in winter, casualties may occur more often in lower speed limit areas, as inclement weather may be associated with worse accidents. Drift is detected through identifying significant changes in classifier performance: usually the classifier's 0–1 loss function when classifying instances. There are multiple approaches to detect such drift [8] such as: sequential analysis, e.g., the Page-Hinkley Test (PHT); statistical process control, e.g., DDM; or monitoring difference between two windows e.g., SEED [11]. A comparison of drift detectors in [3] shows HDDM-W [7] as providing excellent recall with low false-positive drift detection rate. Ensemble frameworks use multiple classifiers which can allow effective classification across different concepts. They usually implicitly handle drift detection by changing which classifiers they use to determine classifications, so do not signal when drift occur. State-of-the-art classification ensembles Leveraging Bagging [4] and Adaptive Random Forest (ARF) [9] have been shown to achieve excellent classification accuracy on commonly-used real world benchmark datasets. Recurring concepts classification frameworks combine a drift detector with a collection of stored classifiers, and generally reintroduce a previous classifier after a drift if it believes a concept has recurred. Diversity Pool (DP) [6] is a state-of-the-art approach that uses entropy measurements to maximise the difference between its pool of classifiers, to improve the chances of having a suitable classifier on drift detection. The Enhanced Concept Profiling Framework (ECPF) [2] trains both a new classifier and a reused classifier from its collection that most accurately classifies recent instances after each drift. When drift next occurs, it retains a copy of the classifier that has classified most accurately since the last drift.

Wang et al. [16] identify major challenges when using explicit drift detection in data streams with class-imbalance: accuracy is disproportionately weighted by the majority class, so a stable majority class may mask change in the minority class; drifts in the minority class may have little impact on accuracy due to classifiers being biased towards correctly classifying the majority class; and classifiers generally take longer to learn the minority class so drops in accuracy can be harder to identify. Some proposed approaches have used different metrics than classification error, *e.g.*, minority class recall, to detect drift but these lead to increased false-positive drift rate [16]. Oversampling, another common solution, affects drift detector guarantees of performance such as that found in HDDM-W [7].

3 Proposed Frameworks

In this section, we first detail our proposed GAD framework and then describe the recurring concepts classification framework, GCPF. We contrast the drift detector framework, GAD, with traditional drift detection to show how it takes steps to address the challenges of explicit drift detection on imbalanced data. GCPF is an adaptation of ECPF that integrates GAD within the framework, improving classification on imbalanced datasets. We provide more detail plus implementations of our frameworks in our code repository at https://github.com/rand079/CrashData.

3.1 GAD: G-Mean Adaptive Drift Detection

GAD tunes the classification threshold of a classifier in a drift detection framework to have a roughly equal chance to detect a positive and negative class correctly. This will allow drifts in the minority class to be detected faster and should improve G-mean. A ROC-curve (*e.g.*, Fig. 1) visualises the trade-off between true positive rate (TPR) and false positive rate (FPR) of a classifier based on its threshold, with the straight dashed line representing expected performance of randomly guessing class labels.

The curve indicates accuracy on each class that a classifier provides when changing the degree of evidence required to classify an example as positive or negative. On imbalanced data, a classifier is trained on more examples of the negative majority class so is more likely to accurately classify this class than the positive minority class. If a classifier misclassifies most minority class instances (*e.g.* point A) then a concept drift affecting that class may not significantly affect overall accuracy and real concept drift may not be detected. Based on recent data, we tune our classification threshold to achieve roughly equal TPR and FPR (*e.g.* point B) by choosing a threshold that achieves the best G-mean on recent data. This should lead to a more obvious drop in accuracy

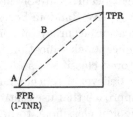

Fig. 1. Classifier ROC curve

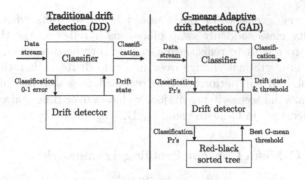

Fig. 2. Traditional drift detection versus GAD

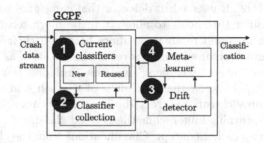

Fig. 3. G-means Concept Profiling Framework

if the classifier becomes less able to correctly classify instances of either class. By changing this threshold, we do not interfere with the classifier training process nor the drift detector's performance guarantees. This should improve G-mean of the framework and ability to detect drift in the minority class.

In traditional drift detection frameworks (DD), as shown in Fig. 2, an instance is classified as it arrives. Once its true class is known, a drift detector determines whether classifier error has significantly increased. If so, drift is signalled, and a new classifier replaces the existing classifier. Classifier thresholds do not vary. In GAD:

1. The classifier classifies each instance as it arrives as determined by its classification threshold.
2. The classifier error passed to the drift detector is based on that threshold and the classifier probabilities that the instance is a positive class is stored in the red-black tree along with the actual class.
3. When drift is detected, GAD uses the classifier probabilities for each instance to calculate the threshold that would have provided the optimum G-mean.
4. A new classifier is created which will use this threshold until it is replaced. The tree is then cleared.

The worst case scenario for GAD would be in datasets where the minority and majority class suddenly swap places, as it relies on the threshold over instances prior to a drift to reflect a reasonable threshold to use after the drift. However, if it is a gradual change over multiple drifts rather than a sudden change, GAD should still perform well as recent data is still relevant. The Electricity benchmark dataset switches majority classes over time and we do not find a substantial decrease in G-mean using GAD.

3.2   GCPF: G-Mean Concept Profiling Framework

We now describe the components within GCPF, and the design of GCPF. GCPF is a classification framework that can learn from each instance as it arrives, adapted from ECPF [2]. It uses a drift detector that can signal when drift occurs. GCPF adapts classifier thresholds to allow it to be more sensitive to drifts in minority classes. It uses and retains classifiers based on their G-mean. GCPF has four major components used during stream classification as shown in Fig. 3:

Classifiers: Two current classifiers, one new and one reused, attempt to learn the relationship between the attributes of a crash and whether a casualty resulted. The **new** classifier is created after drift occurs and is only trained on a drift detector's warning buffer while the **reused** classifier is a copy of a prior classifier trained on an earlier concept. Classifications from the classifier with the highest G-mean are output. *Classifier collection:* This holds a copy of previously trained classifiers. GCPF saves whichever current classifier has highest G-mean when a drift is detected. A classifier is removed if it is found to classify similarly to another by the meta-learner, as GCPF treats the two as representing the same concept: the classifier with the highest G-mean of the two is retained. Classifier similarity is measured on recent data whenever drift is detected. Any of these classifiers may be selected to be reused when drift is detected. *Drift detector:* This monitors the accuracy of the current classifier with the best G-mean. A significant drop in accuracy suggests that concept drift has occurred. Like in GAD, classifier probabilities for each instance are passed to the meta-learner. *Meta-learner:* This calculates the threshold that would have provided the best G-mean over recent data when drift occurs and the next current classifiers use that threshold. The meta-learner also compares behaviour of all classifiers in the classifier collection on data leading up to the drift, and the classifier with the best G-mean on recent data will be reused alongside a new classifier on upcoming data.

4   Experimental Results

In this section, we empirically test that our proposed drift detection framework, GAD, can improve G-mean on synthetic and real-world benchmark imbalanced data streams. We then tune our proposed classification framework, GCPF, on the New Zealand crash dataset, and demonstrate that it can outperform other stream classification frameworks in terms of G-mean on this dataset. We show that it remains fast and memory-efficient in respect to instances seen over large amounts of data, and demonstrate that it generalises, performing well on a classification task with UK crash data.

. Experiments ran on a Intel Core i4470 system with 8GB RAM using MOA v.201706. All tested frameworks use Naïve Bayes classifiers except where specified otherwise. Synthetic dataset generators are available in MOA apart from CIRCLES, which has two numeric attributes in $x, y \in [0, 1]$, with all instances within a circle with centre $(0.5, 0.5)$ and radius r as class 0. CIRCLES, crash datasets, references to other real-world datasets, GAD and GCPF are available in our code repository at https://github.com/rand079/CrashData. All techniques use parameters recommended by authors unless otherwise specified.

NZ Crash Dataset. The New Zealand Transport Authority (NZTA) make past crash data available online. It is updated quarterly and has a detailed data dictionary[1]. The NZTA have provided us a version that orders crashes by date and time, from 2000 up until the end of June 2018. We have added a calculated binary target column which is zero where no casualty occurs and one where casualties have occurred for a crash. All other columns related to casualties have been removed from the dataset. As shown in Fig. 4, the number of crashes in New Zealand increased up to its highest point in 2007 before decreasing, but has increased since 2014. On average, almost 29% of crashes result in a casualty. By year, this proportion ranges from a low of 25% in 2000 to a high of 32% in 2012. There is information on 665,149 crashes reported since Jan 1, 2000. Seventy factors are recorded relating to location, road information, road condition, speed limit and features (such as ditches or fences) involved in the crash. The majority of variables are categorical or binary. As this data is heavily anonymised, date and time of each crash is not provided; nor are any variables relating to the people or vehicles involved.

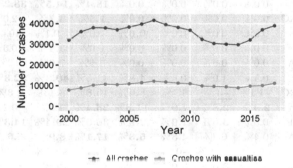

Fig. 4. Number of crashes each year from 2000–2017 in New Zealand

---

[1] https://www.nzta.govt.nz/safety/safety-resources/road-safety-information-and-too lo/disaggregated-crash-data/.

4.1 Experiments Comparing GAD with Traditional Drift Detection

Impact of Drift Detector Choice on G-Mean/FP Drift Rate. Here we examine the impact of using GAD over DD across drift detectors. We test a variety of drift detectors, including two modern, high-performing detectors HDDM-W and RDDM [3]. For these experiments, we used a Hoeffding Tree classifier (using Naïve Bayes at the nodes as per MOA's implementation). As shown in Fig. 5, GAD generally achieves better G-mean than DD, with the only clear exceptions found when using DDM or PHT on Hyperplane. The improvement GAD provides is fairly consistent across drift detectors, with RDDM and SEED notably benefiting more, and PHT benefiting less than other drift detectors. GAD generally appears to achieve better G-mean for more imbalanced datasets. GAD's approach appears to deliver best improvements when there are fewer minority examples for a classifier to learn from.

Table 1. Mean difference in FP drift rate and G-mean when using GAD compared to DD on synthetic datasets with varied imbalance across drift detectors

Metric/Imbal		Diff in FP drift rate				Diff in G-mean			
Detector	Data	0.05	0.1	0.25	0.3	0.05	0.1	0.25	0.33
DDM	Ag	6.4%	0.0%	−0.3%	−3.1%	**30.0%**	**22.5%**	**30.7%**	**19.6%**
DDM	CIRC	1.5%	9.4%	**6.2%**	−4.1%	**8.0%**	**6.7%**	**0.7%**	_−0.9%_
DDM	Hyp	−2.7%	_−9.2%_	−10.8%	_−9.2%_	**8.3%**	**9.0%**	−7.4%	_−26.8%_
DDM	RBF	−3.7%	−1.9%	−1.4%	−1.3%	2.0%	**5.7%**	**1.2%**	0.7%
DDM	SEA	**15.4%**	4.6%	−0.9%	**5.3%**	**6.6%**	**8.1%**	0.5%	_−1.5%_
HDDM-W	Ag	0.0%	0.0%	−0.1%	**1.0%**	4.7%	**9.4%**	**24.8%**	**11.8%**
HDDM-W	CIRC	0.0%	0.0%	0.0%	**0.4%**	0.0%	0.0%	_−2.9%_	_−5.8%_
HDDM-W	Hyp	0.0%	0.0%	**2.3%**	**10.2%**	0.0%	2.1%	**0.4%**	_−0.8%_
HDDM-W	RBF	0.0%	0.0%	−0.3%	**1.2%**	1.2%	**3.3%**	**14.7%**	**5.0%**
HDDM-W	SEA	0.0%	0.4%	**1.0%**	**1.1%**	**6.8%**	**8.7%**	**1.4%**	_−1.0%_
PHT	Ag	0.0%	0.0%	0.0%	0.0%	**18.4%**	**14.5%**	**38.2%**	**25.8%**
PHT	CIRC	0.0%	0.0%	0.0%	0.0%	**1.5%**	**3.7%**	_−2.0%_	_−4.0%_
PHT	Hyp	0.0%	0.0%	0.0%	**0.6%**	0.0%	−0.1%	_−10.4%_	_−8.6%_
PHT	RBF	0.0%	0.0%	0.0%	0.0%	0.0%	**3.4%**	−0.6%	_−6.8%_
PHT	SEA	0.0%	0.0%	0.0%	0.0%	**5.6%**	**7.6%**	0.2%	_−2.2%_
RDDM	Ag	0.5%	1.0%	0.6%	**4.8%**	**41.5%**	**40.3%**	**32.3%**	**18.2%**
RDDM	CIRC	−1.1%	_−3.4%_	1.1%	**8.6%**	**13.8%**	**8.1%**	**2.3%**	−0.2%
RDDM	Hyp	**2.9%**	**2.8%**	**7.9%**	**11.3%**	**26.4%**	1.2%	−0.2%	_−0.7%_
RDDM	RBF	0.5%	**3.0%**	**1.9%**	**2.9%**	**36.0%**	**29.4%**	**14.3%**	**5.2%**
RDDM	SEA	0.4%	0.2%	**4.3%**	**5.3%**	**17.1%**	**8.9%**	0.6%	_−1.8%_
SEED	Ag	−0.6%	−1.2%	−0.2%	−0.6%	**40.1%**	**39.9%**	**34.2%**	**18.3%**
SEED	CIRC	2.1%	**3.0%**	**3.0%**	1.2%	**10.6%**	**12.8%**	**8.0%**	**2.7%**
SEED	Hyp	_−11.3%_	_−6.0%_	_−1.5%_	−0.6%	**30.7%**	**12.8%**	−1.8%	_−2.4%_
SEED	RBF	−0.5%	−2.1%	_−3.7%_	_−5.8%_	**34.7%**	**30.3%**	**14.9%**	**5.0%**
SEED	SEA	0.7%	−1.9%	_−8.8%_	_−8.0%_	**13.3%**	**9.9%**	**2.1%**	_−0.4%_

Bold represents a significantly higher result while underline represents a significantly lower result based on Wilcoxon signed rank tests

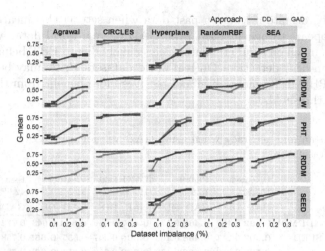

Fig. 5. G-mean for DD (red) and GAD (blue) when run on synthetic datasets with varied imbalance across drift detectors (Color figure online)

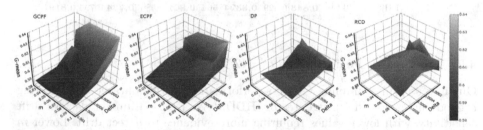

Fig. 6. G-mean with classifier reuse frameworks on NZ Crash data when varying δ and m for HDDM-W

Table 1 quantifies the difference in FP drift rate and G-mean for each drift detector and dataset when using GAD instead of DD. We ran Wilcoxon signed rank tests across experiments to discover significant differences in results, which are represented by bold/underline when GAD got significantly higher/lower results respectively. All drift detectors could provide significantly higher G-mean when using GAD instead of DD. These results support GAD providing larger improvements in G-mean on more imbalanced data.

We tested GAD and DD on real-world benchmark datasets to see if results from synthetic experiments extended to real-world data, with results shown in Table 2. GAD consistently reduced classification accuracy compared to DD, with reductions much more notable when the dataset was more imbalanced. However, DD achieved lower G-mean than GAD on the three datasets with fewer than 20% of instances in the minority class. GAD achieved roughly 0.30–0.55 higher G-mean on these datasets. The reduction in accuracy for GAD on more balanced datasets was much smaller than this. In Electricity, the majority class changes gradually but regularly through the stream. GAD does not classify substantially less accurately, despite relying on past data when setting classification thresholds. PHT appears to benefit most from GAD on real-world data, when it did not benefit as much on synthetic data. This is likely due to PHT being generally conservative in drift detection. On real-world datasets, drifts may be more evident and so PHT drifts more often and receives more benefit from the changed classification thresholds.

Table 2. G-mean and accuracy when using GAD vs. not using GAD on real-world benchmark datasets

Dataset Minority%		UKCrash 15.4%		ClickP 16.8%		PAKDD09 19.7%		Weather 31.3%		Elec 42.5%	
		GAD	DD	GAD	DD	GAD	DD	GAD	DD	GAD	DD
G-mean	HDDM-W	**0.492**	0.162	**0.467**	0.177	**0.497**	0.059	**0.664**	0.648	0.772	**0.846**
	SEED	**0.507**	0.211	**0.468**	0.166	**0.506**	0.067	0.635	**0.655**	0.758	**0.826**
	PHT	**0.500**	0.083	**0.568**	0.170	**0.572**	0.068	0.617	**0.632**	0.795	**0.806**
Accuracy	HDDM-W	0.600	**0.833**	0.382	**0.831**	0.556	**0.801**	0.674	**0.725**	0.765	**0.851**
	SEED	0.511	**0.822**	0.381	**0.832**	0.477	**0.801**	0.643	**0.716**	0.753	**0.835**
	PHT	0.616	**0.843**	0.629	**0.832**	0.564	**0.801**	0.585	**0.734**	0.794	**0.819**

4.2   Framework Performance on NZ Crash Data

Our first step in examining GCPF's performance on this dataset was selecting a good tuning for the drift detector, HDDM-W. Its δ parameter decides drift sensitivity, with lower values requiring more evidence to detect drift. Lower m settings give less weight to recent data while higher m weights recent data more. We tested the degree to which these parameters impacted GCPF's G-mean, along with ECPF, Diversity Pool (DP) and Recurring Drifts Framework (RCD) to see whether GCPF outperformed them. Results are shown in Fig. 6. We used GCPF's optimum parameter settings from our tuning experiment ($\delta = 0.00001$, $m = 0.05$) for the following experiments. We used author-recommended settings for HDDM-W with other drift detection frameworks ($\delta = 0.001$, $m = 0.05$).

We tested multiple approaches to stream classification on the NZ Crash dataset, with results shown in Table 3. GCPF achieves the best G-mean, with reasonably good memory requirements and runtime compared to classifier reuse and ensemble frameworks. GAD achieves the best G-mean apart from GCPF, but gets worse accuracy. Other classifier reuse frameworks do not classify significantly more accurately nor with higher G-mean than one another. The Naïve Bayes classifier, NB, provides a useful baseline for G-mean, and does better than most other frameworks tested. This may be because without drift detection, the classifier learns from more examples of the minority class over time so may be accurate provided large drifts do not occur. We also tested how well GCPF extends to another dataset in a similar area. The UK Crash dataset, as introduced in [1], captures 7 years of road crash data that resulted in casualties in the UK, with 17 categorical variables and one numeric variable. We have

Table 3. G-mean, accuracy, memory use (000s of bytes) and runtime (s) of stream classification frameworks on the NZ Crash data

Type	Framework	G-mean	Accuracy	Memory	Runtime
Classifier	NB	0.598	0.753	**41**	**14.1**
Drift detection	DD	0.585	0.745	49	14.8
	GAD	0.612	0.675	117	17.9
Classifier reuse	ECPF	0.590	0.748	258	20.1
	GCPF	**0.637**	0.713	734	22.3
	RCD	0.595	0.755	1776	15.5
	DP	0.598	0.749	273	34.4
Ensemble	LevBag	0.609	0.752	255	92.7
	ARF	0.544	0.768	251508	275.3

Table 4. G-mean, accuracy, memory use (000s of bytes) and runtime (s) of stream classification frameworks on the UK Crash data

Type	Framework	G-mean	Accuracy	Memory	Runtime
Classifier	NB	0.335	0.817	**34**	**4.4**
Drift detection	DD	0.338	0.807	45	5.5
	GAD	0.433	0.321	66	6.7
Classifier reuse	ECPF	0.235	0.833	89	6.9
	GCPF	**0.497**	0.735	1485	8.2
	RCD	0.334	0.816	8645	142.2
	DP	0.305	0.819	649	10.6
Ensemble	LevBag	0.339	0.815	144	32.7
	ARF	0.104	**0.843**	70423	178.8

altered the class column to be binary, with negative classes representing minor injuries occurring and positive classes featuring severe injuries or fatalities. The dataset has 1 m instances and has 15% of instances being part of the minority class. Table 4 shows results across frameworks on this dataset. We see that GCPF achieves a much higher G-mean than other frameworks. Once again, it underperforms other frameworks in terms of accuracy. Based on these results, our methodology appears to generalise well to a similar problem area.

To test that GCPF would be able to scale well to crash data collected over many years, we needed to check how its memory and runtime requirements grow with larger datasets. To emulate a larger set of data, we ran GCPF over the NZ Crash dataset 10 times in a row (6.7 m instances), so as to plot its time per instance. Figure 7 shows that GCPF maintains a sub-linear time per instance requirement that is much faster than RCD and DP. GCPF has close to linear memory usage per instance. Unlike ECPF, it retains models that are not reused,

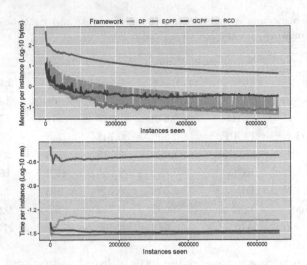

Fig. 7. Runtime and memory-usage per instance for recurring concept frameworks training on NZ Crash data ten times.

so likely grows in size over time. However, after over 6 m instances seen, it still requires substantially less memory than RCD and a similar amount to DP.

5   Conclusion and Future Work

In New Zealand, developments in technology and changes in lifestyle have lead to changes in how roads are used. Traditional analyses of road accidents tend to be retrospective and may not apply to current data. To enable effective drift detection on imbalanced datasets like New Zealand crash data, we have proposed GAD, a new approach to detecting drift in streams with imbalanced classes. We experimentally show using synthetic imbalanced data that GAD provides improved G-mean without severely impacting false-negative drift detection rates. We show that GAD can provide this improvement across drift detectors and classifiers, and that this performance improvement extends to real-world benchmark datasets. Implementing GAD into a recurring concepts framework to create GCPF allows us to outperform even state-of-the-art ensemble classification frameworks in terms of G-mean, while keeping the benefits of explicit drift detection. We show that our technique runs efficiently in terms of time and memory, relative to other recurring concepts frameworks. We show that our proposed approach is generalisable and can extend to a similar problem on UK crash data. Interesting future work could involve extending this technique to data stream classification ensembles. By having a variety of classifier thresholds represented, it could provide a faster way of finding an optimum threshold to maximise G-mean without needing to wait for drift detections to change the threshold. This could allow using the intuition behind GAD on a multi-class classifier.

References

1. Anderson, R., Koh, Y.S., Dobbie, G.: Predicting concept drifts in data streams using metadata clustering. In: IJCNN. IEEE (2018)
2. Anderson, R., Koh, Y.S., Dobbie, G., Bifet, A.: Recurring concept meta-learning for evolving data streams. Expert Syst. Appl. **138**, 112832 (2019)
3. Barros, R.S.M., Santos, S.G.T.C.: A large-scale comparison of concept drift detectors. Inf. Sci. **451**, 348–370 (2018)
4. Bifet, A., Frank, E.: Sentiment knowledge discovery in Twitter streaming data. In: Pfahringer, B., Holmes, G., Hoffmann, A. (eds.) DS 2010. LNCS (LNAI), vol. 6332, pp. 1–15. Springer, Heidelberg (2010). https://doi.org/10.1007/978-3-642-16184-1_1
5. Brzezinski, D., Stefanowski, J.: Prequential AUC: properties of the area under the ROC curve for data streams with concept drift. Knowl. Inf. Syst. **52**(2), 531–562 (2017)
6. Chiu, C.W., Minku, L.L.: Diversity-based pool of models for dealing with recurring concepts. In: IJCNN, pp. 1–8. IEEE (2018)
7. Frías-Blanco, I., del Campo-Ávila, J., Ramos-Jiménez, G., Morales-Bueno, R., Ortiz-Díaz, A., Caballero-Mota, Y.: Online and non-parametric drift detection methods based on Hoeffding's bounds. IEEE TKDE **27**(3), 810–823 (2015)
8. Gama, J., Žliobaitė, I., Bifet, A., Pechenizkiy, M., Bouchachia, A.: A survey on concept drift adaptation. ACM Comput. Surv. **46**(4), 44:1–44:37 (2014)
9. Gomes, H.M., et al.: Adaptive random forests for evolving data stream classification. Mach. Learn. **106**(9–10), 1469–1495 (2017)
10. Gonçalves Jr., P.M., De Barros, R.S.M.: RCD: a recurring concept drift framework. Pattern Recogn. Lett. **34**(9), 1018–1025 (2013)
11. Huang, D.T.J., Koh, Y.S., Dobbie, G., Pears, R.: Detecting volatility shift in data streams. In: 2014 IEEE International Conference on Data Mining, pp. 863–868. IEEE (2014)
12. IRTAD: IRTAD road safety annual report 2018 (2018). https://www.itf-oecd.org/road-safety-annual-report-2018
13. Kubat, M., Matwin, S., et al.: Addressing the curse of imbalanced training sets: one-sided selection. In: ICML, Nashville, USA, vol. 97, pp. 179–186 (1997)
14. Lavrenz, S.M., Vlahogianni, E.I., Gkritza, K., Ke, Y.: Time series modeling in traffic safety research. Accid. Anal. Prev. **117**, 368–380 (2018)
15. Twyford, P., Genter, J.A.: New Zealand government press release: $1.4 billion to save lives on our roads (2018). https://www.beehive.govt.nz/release/14-billion-save-lives-our-roads
16. Wang, S., Minku, L.L., Yao, X.: A systematic study of online class imbalance learning with concept drift. IEEE Trans. Neural Netw. Learn. Syst. **99**, 1–20 (2018)

Interpretability of Machine Learning Solutions in Industrial Decision Engineering

Inna Kolyshkina[1](✉) [iD] and Simeon Simoff[2] [iD]

[1] Analytikk Consulting, Sydney, Australia
inna@analytikk.com
[2] Western Sydney University, Penrith, NSW 2751, Australia
s.simoff@westernsydney.edu.au
http://www.analytikk.com/
https://www.westernsydney.edu.au/

Abstract. The broad application of machine learning (ML) methods and algorithms in diverse range of organisational settings led to the adoption of legislation, like European Union's General Data Protection Regulation, which require firm capabilities to explain algorithmic decisions. Currently in the ML literature there does not seem to be a consensus on the definition of interpretability of a ML solution. Moreover, there is no agreement about the necessary level of interpretability of such solution and on how this level can be determined, measured and achieved. In this article, we provide such definitions based on research as well as our extensive experience of building ML solutions for various organisations across industries. We present CRISP-ML, a detailed step-by-step methodology, that provides guidance on creating the necessary level of interpretability at each stage of the solution building process and is consistent with the best practices of project management in the ML settings. We illustrate the versatility and effortless applicability of CRISP-ML with examples across a variety of industries and types of ML projects.

Keywords: Interpretability in machine learning · Machine learning methodology · Data science methodology · Level of interpretability · Model interpretability · Project management

1 Introduction and Background to the Problem

The rapid increase in the range and diversity of data-driven algorithmic decision engineering has led to the sharp increase of the need for a consistent and comprehensive methodology and process that govern the development, deployment, utilisation and evaluation of the engineering outcomes. Decision engineering, whether in data science (DS) and data analytics (DA) projects or in autonomous systems, like computer-based recommenders and advisers, rely on machine learning (ML) and artificial intelligence (AI) systems, hence, requires

T. D. Le et al. (Eds.): AusDM 2019, CCIS 1127, pp. 156–170, 2019.
https://doi.org/10.1007/978-981-15-1699-3_13

interpretability/explainability of system behavior and decision making outcomes. Interpretability in AI/ML depends on two connected aspects: (i) development of interpretability solutions for AI/ML algorithms and (ii) development of consistent and comprehensive methodology/framework for data science projects, which minimises the risk of project failures and guarantees achieving the necessary (for the project) level of ML/AI system interpretability. Guidotti et al. [13] provide a systematic overview of the current state-of-the-art in (i). Our paper is focused on the development of methodology, which addresses (ii). The rationale supporting such focus is built on the following major arguments: (a) high proportion of data science project failures - an indicator of the need for a consistent and comprehensive methodology and process for ML/AI projects; (b) emerging requirements for sufficient explainability of ML systems - this puts pressure on creation of frameworks/ methodologies which can ensure the sufficient explainability of ML systems; and (c) lack of standard methodology - contemporary methodologies do not include standard consistent components which ensure interpretability through the project. Further we elaborate each of these arguments.

1.1 High Proportion of Data Science Project Failures

Recent reports estimate that between 70% and 85% of data science/ML/AI projects fail. NewVantage survey [1] noted that 77% of businesses see big data and AI initiatives as a big challenge for business. Gartner research [24] argues that 80% of analytics insights will not deliver business outcomes through 2022. McKinsey research [8] reports that 92% of big companies are not successful in using analytics in the organisation. In the past three years the percentage of firms identifying themselves as being data-driven has declined from 37.1% in 2017 to 31.0% in 2019 [1], which is counterintuitive to the expected impact of AI technologies on decision making. Key reasons for these failures are linked to the lack of proper process and methodology in requirements gathering, establishing realistic project timelines, task coordination, communication, and suitable project management framework [1, 29]). Improved methodologies are needed as the existing ones do not cover many important aspects and tasks [17]. Further, studies have shown that the recent biased focus on the tools and systems has limited the ability to gain value from organisational analytic effort [22] and that data science projects need to increase their focus on methodology, including process and task coordination ([12]). Practitioners agree with this view [11].

1.2 Requirements for Sufficient Explainability of ML Systems

In parallel with the above discussed tendencies, there is pressure on creation of frameworks/methodologies, which can ensure the sufficient explainability of the output of the ML systems. Whilst some ML systems (for instance, decision tree and rule induction algorithms) offer methodologically transparent means supporting interpretability/explainability of their output, there is a class of the so-called 'black box' ML models, such as deep neural nets, tree and network

ensembles and support vector machines, which do not provide embedded interpretability. There have been a number of cases where this class of models demonstrated the lack of fairness and poor accuracy [20,25]. In high-stake situations,
systems in which the inner workings are not transparent can be unfair, unreliable, inaccurate and even harmful [6,25]. This view is reflected in legislations like
the European Union's General Data Protection Regulation (GDPR) [2], though
there are also warnings to policymakers to be aware of potential impact of legislations like GDPR on AI and emerging algorithmic economy. These developments increase the pressure on creation of frameworks and methodologies, which
can ensure the sufficient explainability of AI and ML solutions. A report by AI
Now Institute [23] recommended standardising the AI and ML system-building
process and incorporating relevant algorithmic impact assessments into the processes the organisations already use. Many organisations and major technology
developers are following this recommendation [4].

1.3   The Lack of Standard Methodology

Though having a good methodology is important for the project success, so far
there is no formal standard for methodology in the data science projects [26].
CRISP-DM methodology [28], created in the late 1990s, is considered the de-
facto standard [5,14]. It is industry-, tool- and application agnostic [17]. It is not
fully meeting the needs of data science community, and its usage appears to be
decreasing [26]. While various extensions of the methodology, including IBM's
ASUM-DM and Microsoft's TDSP were proposed, at this stage none of them has
become the standard. Many CRISP-DM extensions are fragmented and either
propose additional elements into the data analysis process, or focus on organisational aspects without the necessary integration of domain-related factors [21].
Finally, while methodologies from related fields, like the agile approach used in
software engineering, are being considered for use in data science projects, there
is no full clarity on whether they are fully suitable for the purpose [15], therefore
we did not include them in the scope of this paper.

1.4   Opportunities in Creating Interpretability-Related
Methodologies

Recent state-of- the-art reviews related to interpretability [7,9,19] as well as more
algorithm-focussed reviews [13,16,18] report that: (i) interpretability of AI and
ML solutions and the underlying models is not well defined; (ii) the work related
to interpretability is scattered throughout a number disciplines, including AI,
ML, human-computer interaction (HCI), visualisation, cognition; and (iii) current research seems to address a particular category or technique instead of the
overall concept of interpretability. Similarly, while there is a number of suggested
approaches to measuring interpretability [18], there is no consensus neither on
measuring or evaluating the *level of interpretability* nor on the best type of explanation metric [9]. Currently there is confusion about the interpretability notion
[19], including a lack of clarity about how the many proposed interpretation

approaches can be evaluated and compared against each other, how to choose a suitable interpretation method for a given business issue and audience as well as limited guidance on how interpretability can actually be used in data-science life cycles. The lack of consensus gives an opportunity to create a comprehensive methodology, which takes into account different perspectives and aspects of interpretability (comprehensibility), such as predictive accuracy, bias, noise, sensitivity, faithfulness, specificity, local interpretability, global interpretability and domain specifics.

2   Methodology of Establishing and Building the Necessary Level of Interpretability of ML Business Solution

2.1   The Necessary Level of Interpretability of an ML Solution

In line with interpretability in Google's responsible AI practices [4] and expanding on [10] approach, we introduce the concept of *necessary level of interpretability* (NLI) of a business ML solution as the combination of the degree of accuracy of the underlying algorithm and the extent of understanding of the inputs, inner workings, the outputs, the user interface and the deployment aspects of the ML solution that is required to achieve the project goals. If this level is not achieved, the solution will be inadequate for the purpose. This level needs to be established and documented at the initiation stage of the project as part of requirements collection. We then describe a ML system as *sufficiently interpretable or not* based on whether or not it achieved the required level of interpretability.

Obviously, this level will differ from one project to another depending on the business goals and agreed measures of interpretability. If individuals are directly and strongly affected by the solution-driven decision - for example, in medical diagnostic or legal settings - then both the ability to understand and trust the internal logic of the model as well as the ability of the solution to explain individual predictions are extremely important. In other cases, when a ML solution is used in order to inform business decisions about policy, strategy or interventions aimed to improve the business outcome of interest, then it is the need to understand and trust the internal logic of the model that is of most value and individual predictions are not the focus of the stakeholders. For example, in one of our projects an Australian state organisation wished to establish what factors influenced the proportion of children with developmental issues and what interventions can be undertaken in specific areas of the state in order to reduce that proportion. The historical, socioeconomic and geographic data provided for the project was aggregated at a geographic level of high granularity.

In other cases, for example, in the case of an online purchase recommender solution, the overall outcome, such as increase in sales volume, may be of higher importance, than the interpretability of the model. Similar requirements of solution interpretability were in a project where an organisation owned assets that were located in remote areas and were often damaged by birds or animals nests.

The organisation wished to lower their maintenance cost and planning by identifying as soon as possible the assets where such nests were present instead of doing expensive examination of each asset. This was achieved by building a ML solution that classified Google Earth images of the assets into those with and without nests. In this project it was important to identify as accurately as possible a proportion of assets with nests on them, while misclassifying an individual asset image was not of great concern.

2.2 CRISP-ML Methodology

The proposed methodology of building interpretability of a ML system is based on our methodology CRISP-ML. It is an updated version of CRISP-DM and is industry-, tool- and application-agnostic. It seamlessly accommodates modern ML techniques and creates the NLI through the whole ML solution creation process. In order to explain how to ensure that the NLI of ML system is achieved in the project, we elaborate its seven stages, summarised in Fig. 1. We illustrate key concepts with real-world examples/mini case studies.

The Project Initiation and Planning Stage. Interpretability Matrix

Objectives and Importance. This stage is crucial for the overall project success [3] and for the system interpretability building. It covers *the activities needed to start up the project*, including (i) the identification of project sponsor/key stakeholders and preparation of project charter – a document that outlines project objectives, scope, high-level deliverables, assumptions, constraints, and risks – after being signed off it serves as a reference of authority for the future; and (ii) the *planning activities* such as collecting requirements; agreeing upon initial data to use; preparing a detailed scope statement; estimating effort, duration and costs; assessing and responding to risks; developing communications documents, project schedule and plan and finally, obtaining project sponsor's approval to proceed with the project.

Establishing the Necessary Level of Interpretability. NLI is established as part of requirements collection. It is driven by the project objectives, and also influenced by domain specifics, stakeholder requirements, project constraints, industry regulator requirements to name the key factors. Proper requirements collection (i.e. determining and documenting conditions or tasks that must be completed to meet the project objectives) is crucial to the project success [3]. As part of requirement gathering we work with key stakeholders to determine NLI of the solution. Typically, this may require that the relevant stakeholders have a clear understanding of the (i) *data inputs used* - are they reliable, of suitable quality and representative of the real-world data; (ii) *solution outputs* - are they consistent with the project goals in terms of accuracy, format, ease of understanding for the end users, level of potential business insight, and are they valid from the ML and business perspectives; (iii) *format* they should be provided to the

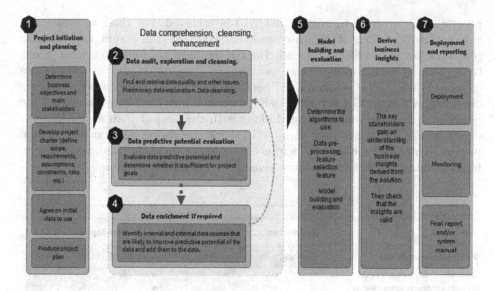

Fig. 1. Conceptual framework of CRISP-ML methodology

end user, e.g. tables, visualisations, graphs, infographics and other representations; (iv) *high-level modelling approach*, its validity and whether it is proven and likely to work in this industry; (v) *implementation process* of the solution in the organisational systems, and how it should be audited, monitored and updated.

For example, in a project in workers compensation insurance that aimed to identify cases likely to become expensive, the objectives included building a ML system that would: (i) explain what factors and to what extent were influencing the outcome of interest i.e. claims cost; (ii) allow the organisation to derive business insights that will help make data-driven accurate decisions regarding what changes can be done to improve the outcome i.e. reduce the likelihood of an expensive claim by a specified percentage; (iii) be accurate, robust and able to work with real-world organisational data; and (iv) have easy-to-understand outputs that would make sense to the executive team and end users (case managers) and that the end users could trust. The established interpretability requirements in this project included: (i) having trustworthy, quality data inputs, representative of the organisational data that the solution would be deployed on; (ii) the outputs should be provided as business rules that were were easy to understand for end users and to deploy on organisational data; (iii) the high-level algorithmic approach had to be easily understood by the executive team and the BI team who would monitor its performance; (iv) explain at least 80% of variation in the data, be valid from the ML point of view; and (v) its outputs needed to make sense to the domain experts.

Creating Project Interpretability Matrix as Part of Requirements Collection. The next step is to establish what needs to be done by each stakeholder at each project stage in order to ensure that NLI is achieved. For this, we create the *interpretability matrix* (IM), whose *rows* show CRISP-ML stages, and *columns* represent key stakeholders. In each IM cell we need to document what needs to be done by each stakeholder at each project stage to ensure that NLI of the solution is achieved. Completed IM becomes part of the business requirements document; the activities it outlines are integrated into the project plan and are performed, updated and monitored along with the project plan as needed. For example, Fig. 2 shows a very high-level IM for the above mentioned insurance project. The green, yellow and white colour background indicate, respectively, high, medium and low level of involvement of a stakeholder group.

| Project stages | Key Stakeholders | | | |
	Executive team	DE team	IT team	Modelling team
Project initiation and planning	Establishing project requirements, understanding domain aspects; agreeing on the initial data to use			
Data audit. Quality and sanity checks	Low	clarify domain-related aspects and provide sanity checks	Low	Data analysis; Data predictive potential evaluation and enrichment. Algorithm development and evaluation
Evaluation of data predictive potential	Low		Low	
Data enrichment if required	Low		Low	
Model building and evaluation	Low		Low	
Deriving business insights	Evaluating business insights from modelling team's report		Low	Preparing a report and a presentation for ET
Deploying. Monitoring. Updating	Low		Deployment	Preparing a solution manual

Fig. 2. Example of a very high-level interpretability matrix for the insurance project. (Color figure online)

Entries to the Interpretability Matrix at Each Stage of CRISP-ML

Further we discuss typical entries to the interpretability matrix (IM) at each stage of CRISP-ML, and illustrate them with real-world examples. Usually, in our experience key stakeholders for ML system projects are the executive team (ET); the data provider (DP) team which is often a part of the organisational IT team; the domain experts (DE) and the modelling team (M). These abbreviations will be used in the below descriptions along the stages IM.

Stage 1. Figure 3 provides details of IM content related to this stage.

Stages 2–4. Stages 2, 3 and 4 in Fig. 1 are mainly data-related and form the data comprehension, cleansing and enhancement mega-stage. Further we consider the content of interpretability matrix for each individual stage.

	ET	DE	IT/DP/Deployers	M		
Stage 1. Project initiation and planning	Stage aim: Identify project sponsor and key stakeholders, prepare the project charter; collect requirements; agree upon initial data to use; prepare a detailed scope statement; estimate effort, duration and costs; do risk analysis; develop communication plan etc.					
	Work together with the other stakeholders to establish and/or gain in-depth understanding of the project objectives; scope; assumptions, etc. Gather requirements, specifically establish and document the required interpretability level including the minimal required accuracy and understanding of inputs, outputs, inner workings and implementation aspects.					
	Gain a high-level understanding of the approach for the ML solution building process and potential outcomes. Help modellers to gather requirements, establish assumptions, constraints etc.			Gain in-depth understanding of requirements and interpretability level Gain a high-level understanding of a) domain aspects relevant to the project goals achievement; b) the data agreed to be used.		

Fig. 3. CRISP-ML: Stage 1 - typical IM content related to this stage.

Stage 2. Data audit, exploration and cleansing play a key role in the development of stakeholder trust in the approach and ultimately in the solution, if achieving user trust in the solution is part of the established NLI for the project. Figure 4 demonstrates the typical content of IM at this stage. This stage is important in any project where interpretability is of high priority, because wrong data values may slide in unnoticed and skew the outputs. For example, in a project aiming to establish what drives morbidity of pregnant women with diabetes and their children, the data on the age of the mother had records of 99 yo. Domain experts clarified that '99' was a code for 'Age Unknown'.

Stage 3. Figure 5 demonstrates the typical content of the interpretability matrix related to the evaluation of the predictive potential of the data. This stage is often either omitted or not stated explicitly in other processes/frameworks (for example, in CRISP-DM), however it is crucial in terms of achieving NLI because it establishes whether the information in the data is sufficient for achieving the project goals (for example, for explaining the outcome of interest). At this stage, in-depth data exploration and preliminary modelling is performed, where several advanced and powerful non-supervised and supervised ML techniques are used to explore the data, establish the most promising strategies for feature engineering/data transformations and modelling and assess whether the initially identified data and other resources are sufficient for achieving the business goals. The choice of the ML techniques is tailored for each project; detailed description of them and the process of assessment of the predictive potential is beyond the scope of this paper. Techniques used for estimating predictive potential include

Stage 2. Data audit, exploration and cleansing.	**Stage aim:** Demonstrate that the data inputs into the solution are of adequate quality, make sense to DEs and are representative of the real-world data that the solution will be deployed on.			
	ET	DE	IT/DP/Deployers	M
	Low	Provide domain-related knowledge to help modellers to develop the needed data understanding, conduct data quality assessment and run relevant data "sanity checks".	Provide data-related knowledge to help modellers to develop the needed data understanding and conduct data quality assessment.	Develop the needed data understanding, conduct data quality assessment and run relevant data "sanity checks".

Fig. 4. CRISP-ML: Stage 2 - typical IM content related to this stage.

components of various dimensionality reduction approaches, advanced clustering methods and proven highly-predictive methods such as random forest, boosting methods and deep neural networks).

In our experience, initially identified data often needs to be enriched by external data. For example, in the insurance example the predictive potential of the data containing claim and worker data history was shown to be insufficient for the project objectives. The domain experts suggested to enrich the initial data with the history of what doctors and other health service providers a worker saw, the medicines worker was prescribed (for example, opioids) and some other data. Adding these data significantly improved the model accuracy. Enrichment by additional data is not always needed. Specifically, from our experience, image and free text data often do not require additional information to build an accurate model. For example, in a project where social media data were used to compare customer perception of the four Australian major banks, at stage 3 we established that collected data were enough for project purposes, but additional in-depth feature engineering was required.

Stage 3. Data predictive potential evaluation	**Stage aim:** assess whether the initially identified data is sufficient for project purposes.			
	ET	DE	IT/DP/ Deployers	M
	Low	Consultation and clarification of any domain-related aspects	Low	Assess data predictive potential using ML methods

Fig. 5. CRISP-ML: Stage 3 - typical IM content related to this stage.

Stage 4. Figure 6 shows a typical content of IM if it was determined at stage 3 that the initially identified data or other resources are not sufficient for the project purposes and the data have to be enriched. In practice this involves additional analysis, usually, data enrichment by adding new data, less often by in-depth feature engineering of the existing data. Additional internal and external data sources are identified, the new data is extracted, audited, cleansed and added to the previously used data. Then predictive potential of the enriched data is again assessed by applying the same ML methods as in stage 3.

Stage 4. Data enrichment	Stage aim: additional internal and external data sources are identified, the new data is extracted, audited, cleansed and added to the previously used data. Then predictive potential of the enriched data is again assessed. This step is repeated until the necessary level of predictive potential is achieved.			
	ET	DE	IT/DP/Deployers	M
	Low	Help identify the data sources suitable for enrichment and clarify any domain-related aspects	Help identify the data sources suitable for enrichment and clarify any data-related aspects	Access the new data, conduct data enrichment and return to stage 3 to evaluate whether the enriched data is sufficient for the project purposes

Fig. 6. CRISP-ML: Stage 4 - typical IM content when data enrichment is required.

This step is repeated until the necessary level of predictive potential is achieved or, if it has been established that achieving it is impossible, this finding is further discussed with the key stakeholders and the relevant decisions are made. Thorough planning at Stage 1 minimises the risk of that occurring. In the insurance example described above, data enrichment was a key step. The fact that the model showed that the cost of a claim can be significantly dependent on the providers a worker visited, built further trust in the solution because it confirmed the domain experts hunch that they previously had not had enough evidence to prove.

Stage 5. Figure 7 shows a typical content of IM for the model building and evaluation stage. To achieve NLI, modellers have to choose the appropriate technique(s) that will balance the required outcome interpretability with the required accuracy and with other requirements/constraints (e.g. the needed functional form of the model and/or algorithm). The ML techniques to be used for modelling are selected taking into account the predictive power of the model, its suitability for the domain and the task, and NLI. The data is pre-processed, and modelled and model performance is evaluated. Detailed description of the process of algorithm choice and model assessment is beyond the scope of this paper and will be covered in a separate publication.

Stage 5. Model building and evaluation	Stage aim: select ML techniques to be used for modelling; preprocess the data, build the model/models; evaluate model performance/choose the best model			
	ET	DE	IT/DP/Deployers	M
	Low	Clarify any domain-related aspects	Low	Select the ML techniques to be used for modelling taking into account interpretability requirements; preprocess the data (including feature selection and feature engineering); build the model and evaluate its performance. Choose the model whose performance level is as required by the necessary interpretability level established at stage 1.

Fig. 7. CRISP-ML: Stage 5 - typical IM content indicating how NLI influences the strategy of choosing modelling techniques by the modelling team.

In the insurance example, the solution output had to be produced in the form of business rules. Therefore the feature engineering methods and modelling algorithms used included rule-based techniques such as decision trees and association rule-based methods. In another example, a large Australian asset-owning organisation needed a ML solution that would help them to proactively optimise asset maintenance planning and cost and asset failure risk reduction as well as to justify funding requests to industry regulator. The regulator specifically requested that the solution be delivered in the linear model form. Such a requirement towards the model type is common in some areas. For example, in credit risk assessment certain models have to be in the logistic regression format. Often there is no constraint on the model functional form. For example, in the above mentioned image classification project, we simply used the most accurate model we could build, which turned out to be a convolutional neural network. Some other techniques used in stage 5 include boosted regression trees, random forest, LASSO methods and deep neural networks.

Stage 6. Figure 8 shows how the IM reflects the role of interpretability in the formulation of business insights necessary to achieve the project goals and in helping the ET and end users to understand the derived business insights and to develop trust in them. DE team might also have a medium to low level of involvement for clarification of any domain-related aspects.

For example, in the insurance project modellers and DEs prepared a detailed presentation for the ET explaining not only the learnings from the solution but also the high-level model structure and its accuracy. In the image processing project, on the other hand, the presentation was focussed on the results and their accuracy rather than on model inner workings.

	Stage aim: Executives (and end users if relevant) gain an understanding of the business insights derived from the solution, develop an understanding that the insights are valid and valuable and build trust in these insights.			
	ET	DE	IT/DP /Deployers	M
Stage 6. Deriving business insights.	Gain an understanding of the business insights derived from the solution, develop an understanding that the insights are valid and valuable and build trust in these insights.	Clarification of any domain-related aspects	Low	Help the executives and end users to develop the trust and understanding. Usually, they prepare presentations or reports to achieve that.

Fig. 8. CRISP-ML: Stage 6 - typical content of IM related to this stage.

Stage 7. Figure 9 shows the shift of responsibilities for ensuring the achieved interpretability level is maintained during the future use of the solution. At this stage, a deployment is conducted if required and monitoring/updating process and schedule is prepared, based on the developed technical report.

This stage and the related interpretability aspects differ significantly depending on the project goals. We illustrate this diversity with some of our projects. In the childhood development project no deployment was required, but a report and a visualisation of the solution was needed. In the nest identifying project no deployment was required, but the list of assets likely to have nests on them was needed as well as a brief report and the model code. In the insurance and

	Stage aim: Deployment is conducted if required and monitoring/updating process and schedule is prepared. A technical report is created.			
	ET	DE	IT/DP/ Deployers	M
Stage 7. Deployment and reporting.	Low	Clarification of any domain-related aspects	Gain understanding of any information regarding the solution that is necessary for its deployment and maintenance. Deployment and related activities	Write technical report, manual or other documentation as required explaining the aspects needed to ensure the necessary interpretability level Provide consultation about model specifics to the deployer team

Fig. 9. CRISP-ML: Stage 7 - activities ensuring the achieved interpretability level is maintained during the future utilisation of the solution.

asset management examples deployment was needed, as well as a full technical report, a solution manual and an updating and monitoring recommendations.

2.3 Conclusions

This article addresses the problem of providing companies with capabilities to explain algorithmic decision engineering. We introduced a definition of interpretability of an end-to-end business ML solution, the necessary level of interpretability of such solution and a methodology (CRISP-ML) of achieving it. CRISP-ML integrates interpretability aspect into the overall framework instead of just at the modelling stage. It requires to take more than the algorithm accuracy into consideration when deciding what the 'best' model is by pushing questions about use and interpretability up front. Further, it defines the responsibilities of different stakeholders to ensure that this is done.

CRISP-ML is an extension of CRISP-DM, which enables organisations to (i) establish shared understanding across all key stakeholders about the solution and its use; (ii) build stakeholder trust in the solution outputs; and (iii) get buy-in from all relevant parts of the organisation. It allows the end users to confidently interpret the solution results and make successful evidence-based business decisions. If needed, they can explain these decisions to any external party. We successfully applied this methodology in commercial projects across a variety of industries including banking, insurance, utilities, retail, FMCG, public health and transport to name some areas. It effortlessly accommodates the diversity of industry specifics as well as variety of organisational goals, ML techniques and data types. While comparing the effectiveness of this methodology to other approaches is beyond the scope of this paper, future work includes experimental assessment similar to the one performed in [27].

References

1. Big Data and AI executive survey. Technical report, NewVantagePartners LLC (2019)
2. General Data Protection Regulation (GDPR). Official Journal of the European Union L 119/1 (2016). https://gdpr-info.eu/
3. PMBOK® Guide - Sixth Edition. Project Magament Institute (2017)
4. Google AI: Responsible AI Practices - Interpretability. https://ai.google/responsibilities/responsible-ai-practices/?category=interpretability. Accessed 5 Aug 2019
5. Ahmed, B., Dannhauser, T., Philip, N.: A Lean Design Thinking Methodology (LDTM) for machine learning and modern data projects. In: Proceedings of 2018 10th Computer Science and Electronic Engineering (CEEC), pp. 11–14. IEEE (2018)
6. Dawson, D., et al.: Artificial intelligence: Australia's ethics framework. Technical report, Data61 CSIRO, Australia (2019)
7. Doshi-Velez, F., Kim, B.: Towards a rigorous science of interpretable machine learning. arXiv e-prints arXiv:1702.08608, February 2017

8. Fleming, O., Fountaine, T., Henke, N., Saleh, T.: Ten red flags signaling your analytics program will fail. Technical report, McKinsey & Company (2018)
9. Gilpin, L.H., Testart, C., Fruchter, N., Adebayo, J.: Explaining explanations to society. CoRR abs/1901.06560 (2019). http://arxiv.org/abs/1901.06560
10. Gleicher, M.: A framework for considering comprehensibility in modeling. Big Data 4(2), 75–88 (2016)
11. Goodson, M.: Reasons why data projects fail. KDnuggets, November 2016. https://www.kdnuggets.com/2016/11/ten-ways-data-project-fail.html
12. Grady, N.W., Underwood, M., Roy, A., Chang, W.L.: Big data: challenges, practices and technologies: In: NIST Big Data Public Working Group workshop at IEEE Big Data 2014. Proceedings of IEEE International Conference on Big Data 2014, pp. 11–15 (2014)
13. Guidotti, R., Monreale, A., Ruggieri, S., Turini, F., Giannotti, F., Pedreschi, D.: A survey of methods for explaining black box models. ACM Comput. Surv. 51(5), 93:1–93:42 (2018)
14. Huang, W., McGregor, C., James, A.: A comprehensive framework design for continuous quality improvement within the neonatal intensive care unit: integration of the SPOE, CRISP-DM and PaJMa models. In: Proceedings of IEEE-EMBS International Conference on Biomedical and Health Informatics (BHI), pp. 289–292 (2014)
15. Larson, D., Chang, V.: A review and future direction of agile, business intelligence, analytics and data science. Int. J. Inf. Manag. 36(5), 700–710 (2016)
16. Lipton, Z.C.: The mythos of model interpretability. ACM Queue 16(3), 30:31–30:57 (2018)
17. Mariscal, G., Marbán, O., Fernández, C.: A survey of data mining and knowledge discovery process models and methodologies. Knowl. Eng. Rev. 25(2), 137–166 (2010)
18. Molnar, C., Casalicchio, G., Bischl, B.: Quantifying interpretability of arbitrary machine learning models through functional decomposition. arXiv:1904.03867
19. Murdoch, W.J., Singh, C., Kumbier, K., Abbasi-Asl, R., Yu, B.: Interpretable machine learning: definitions, methods, and applications. arXiv:1901.04592
20. O'Neil, C.: Weapons of Math Destruction: How Big Data Increases Inequality and Threatens Democracy. Crown Publishers, New York (2016)
21. Plotnikova, V.: Towards a data mining methodology for the banking domain. In: Kirikova, M., et al. (ed.) Proceedings of the Doctoral Consortium Papers Presented at the 30th International Conference on Advanced Information Systems Engineering, CAiSE 2018, pp. 46–54 (2018)
22. Ransbotham, S., Kiron, D., Prentice, P.K.: Minding the analytics gap. MIT Sloan Manag. Rev. 56, 1 (2015)
23. Reisman, D., Schultz, J., Crawford, K., Whittaker, M.: Algorithmic impact assessments: a practical framework for public agency accountability. Technical report, AI Now Institute, April 2018
24. Roy Schulte, W., et al.: Predicts 2019: data and analytics strategy. Technical report, Gartner Research, November 2018
25. Rudin, C.: Stop explaining black box machine learning models for high stakes decisions and use interpretable models instead. Nat. Mach. Intell. 1, 206–215 (2019)
26. Saltz, J.S., Shamshurin, I.: Big data team process methodologies: a literature review and the identification of key factors for a project's success. In: Proceedings of 2016 IEEE International Conference on Big Data 2016, pp. 2872–2879 (2016)
27. Saltz, J.S., Shamshurin, I., Crowston, K.: Comparing data science project management methodologies via a controlled experiment. In: HICSS (2017)

28. Shearer, C.: The CRISP-DM model: the new blueprint for data mining. J. Data Warehouse. **5**, 13–22 (2000)
29. Stieglitz, C.: Beginning at the end - requirements gathering lessons from a flowchart junkie. In: PMI® Global Congress 2012–North America, Vancouver, British Columbia, Canada. Project Management Institute, Newtown Square, PA (2012)

Customer Wallet Share Estimation for Manufacturers Based on Transaction Data

Xiang Li[1,2]([✉]), Ali Shemshadi[2], Lukasz P. Olech[2,3],
and Zbigniew Michalewicz[1,2,4,5]

[1] School of Computer Science, University of Adelaide, Adelaide, SA 5005, Australia
[2] Complexica Pty Ltd., 155 Brebner Drive, West Lakes, SA 5021, Australia
{xl,as,lo,zm}@complexica.com
[3] Wroclaw University of Science and Technology, Wybrzeze Stanislawa
Wyspianskiego 27, 50-370 Wroclaw, Poland
[4] Institute of Computer Science, Polish Academy of Sciences,
Ordona 21, 01-237 Warsaw, Poland
[5] Polish-Japanese Academy of Information Technology,
Koszykowa 86, 02-008 Warsaw, Poland
http://www.adelaide.edu.au
http://www.complexica.com
https://pwr.edu.pl/en/
https://ipipan.waw.pl/en/
https://www.pja.edu.pl/en/

Abstract. The value of customers for any business cannot be over-emphasised, and it is crucial for companies to develop a good understanding of their customer base. One of the most important pieces of information is to estimate *the share of wallet* for each individual customer. In the literature a related concept is often referred to as *customer equity* that provides aggregated measures such as the business market share. The current trend in personalising marketing campaigns have led to more granular estimation of wallet share, than the entire customer base or aggregated segments of customers. The current trend in personalising marketing and business strategies have lead to more granular estimation of wallet shares than the entire customer base or aggregated segments of customers. Existing research in this area requires access to additional information about customers, often collected via various surveys. However, in many real-world scenarios, there are circumstances where survey data are unavailable or unreliable. In this paper, we present a new customer wallet share estimation approach. In the proposed approach, a predictive model based on decision trees facilitates an accurate estimation of wallet shares for customers relying only on transaction data. We have evaluated our approach using real-world datasets from two businesses from different industries.

Keywords: Wallet share estimation · Customer equity · Random Forest · Real-world case study

© Springer Nature Singapore Pte Ltd. 2019
T. D. Le et al. (Eds.): AusDM 2019, CCIS 1127, pp. 171–182, 2019.
https://doi.org/10.1007/978-981-15-1699-3_14

1 Introduction

Wallet share can be defined as the ratio of money that a customer spends with a brand compared to all of his/her expenditure on similar brands that can be considered as competitors. It can help businesses to understand and evaluate their relationship with their customers [7]. Thus it is crucial for businesses to have the ability to measure the share of wallet of their customers. Different businesses, including manufacturers and distributors, usually record a significant amount of data on their customers.

One of the main approaches to estimating share of wallet is to use Voice of Customer (VoC) data [13]. VoC data can be collected by sales representatives or call centres either by surveying customers (pull) or by monitoring the messages sent from customers when they initiate. Many industries such as retailers [3] or banks [2] have access to an extensive amount of VoC data. However, this is not the case manufacturers, which usually do not possess a well-sized sample of VoC data compared to other industries, even when they also operate as distributors.

Research in this area has identified different requirements in real-world scenarios of wallet share estimation. Thus different trends can be observed in the literature including "the analysis of customer wallet share and its impacts in different environments" and "the development of new approaches to estimate the share of wallet for customers based on the availability of the data". In this paper, our focus is more on the latter while working on a novel application area (manufacturers) from the former trend's perspective.

Given that a manufacturer has access to transaction data only through retailers and has a limited amount of VoC data through the sale process, we investigate a novel approach for measuring the share of wallet for manufacturers. The main research questions, which we aim to answer in this paper, are as follows:

1. How is it possible to accurately measure the share of wallet for individual customers for a given manufacturer based only on transaction data?
2. What are the most important features of transaction data to build an effective predictive model of wallet share?
3. Can decision tree-based modelling be deployed to facilitate a real-life predictive model for wallet share estimation?

Figure 1 illustrates our scenario. In this paper, we focus on manufacturers that distribute products to their customers via a chain of retail shops. Dashed lines denote the flow of material, and solid lines denote the flow of information.

Occasionally, a manufacturer may contact its customers for a survey, however, the information collected usually does not represent the general population due to its limited sample size. In order for the manufacturer to optimise its offers and promotions in different areas, it needs to estimate its wallet share with each customer. In our scenario, a decision tree-based model is developed to estimate the wallet share solely through transaction data that have been collected through retail shops.

The rest of this paper is organised as follows. First, an overview of the related research is presented in Sect. 2. Section 3 describes the proposed approach to

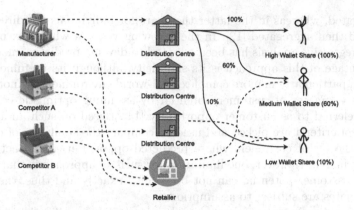

Fig. 1. Conceptual illustration of wallet share scenario

estimate wallet share estimations based on transaction data for manufactures. Details of the evaluation results and experiments on two real-world datasets are presented in Sect. 4. Section 5 concludes the paper and provides some direction for future research.

2   Related Work

In this section, we review the research related to our work. Wallet share estimation is a fundamental problem that has been investigated by many researchers for more than a decade [6]. Different methods have been proposed by different researchers to tackle the challenges present in this area.

In the literature, two main research trends are identified. The first trend focuses on the analysis of customer wallet share and its impacts in different environments. At the highest level, the impact of wallet share has been assessed as crucial in both Business to Business (B2B) as well as Business to Customer (B2C) [6] environments. At the lower level, research narrows it down to particular business domains (e.g., retail, wholesale, manufacture).

In the domain of retailing, research has demonstrated a relationship between customer satisfaction and the share of wallet [10]. While this relationship has been described as positive, yet it is being considered relatively weak. However, research in some particular areas of retailing (e.g., retail banking, mass-merchant retail, and Internet service providers) shows a more sophisticated correlation between the share of wallet, customer satisfaction and other business goals [1,5].

The second trend (which is more relevant to the research presented in this paper) focuses on new approaches for estimating the share of wallet for customers when provided with a different level of available data. Distinguishing criteria in this area are the prediction (estimation) approach, as well as the input data.

The research prior to 2005 considered two main approaches for wallet share estimation: top-down and bottom-up. In the former one, the market share is

dis-aggregated, whereas in the latter the share for individuals are directly estimated and then aggregated [12]. In the following years, a white-box modelling
based on regressive analysis has been adopted to develop predictive models [11].
The advantage of this approach is its simplicity, although its usefulness would
be limited, particularly in more complex, real-world environments. Another approach uses estimates of customer potential by assuming optimistic conditions,
which is referred to as customer opportunity [14]. Based on such an approach,
a number of criteria are picked and used to compare the similarity of each customer to the set of customers whose predicted opportunities and actual sales
are a close match. However, one limitation for such an approach is that in many
cases, the customer potential can not be verified exactly and thus, the validity
of all estimates are subject to assumptions.

Finally, addressing requirements based on input data specifications is a trend-
ing field of research in oCthis area—and this is directly related to this paper.
Many papers have addressed a number of different requirements. In particular, the use of V data is one of the main approaches to estimate the share of
wallet [13]. VoC data can be collected using survey data or unstructured data
gathered. However, in many circumstances, VoC data are not available. In this
case, existing approaches have tackled data availability issues by utilising transaction data for credit cards and focusing on inter-purchase times [2]. However,
unlike manufacturers, credit card companies usually hold many records on their
customers' transactions. Other research, which addresses the same issue, do not
rely on real-world data and need to be further extended to include only the
transaction data [4].

3   Methodology

This section starts with a brief introduction of the available raw data (Sect. 3.1).
Then Sect. 3.2 presents the set of features extracted from the raw data. Estimation algorithm details are reported in Sect. 3.3.

3.1   Datasets

In this research, two datasets from different Australian manufacturers are used.
The first one came from the paint industry and the second one—from a major
producer of air conditioning products. In both cases, the companies behave both
as a manufacturer and a distributor. As distributors, they have performed surveys on selected customers. But the coverage of these surveys was limited and
skewed towards positive feedback. In our research, we have used the existing
survey results as the training input to build the estimation models.

For each dataset, transaction data up to 3 years and around 250 customers
are available. Customers are usually small or medium-sized enterprises, e.g.,
builders, handymen, electricians or painters. The data are extracted directly from
the manufacturer's sales database. The transaction data contained: *Order Date*,
Customer ID, *Product ID*, *Product Quantity*, and *Product Price*. The response

variable is an integer *walletshare* ranging from 0 to 100. 0 means the customer doesn't buy anything, and 100 means the customer buys everything needed from this manufacturer. The value was provided by sales representatives who had high confidence in the reported score. This confidence arises from a number of reasons, e.g., long-lasting cooperation, built trust. No data from the survey is used except the response variable *walletshare*.

3.2   Factorisation

The raw data is factorised into a row-based matrix, where each row represents one customer. The columns of the matrix are the features extracted from the raw data. The list has been created after extended discussions with domain experts. The main intuition behind those features is that high wallet share customers should buy products in a more consistent way than low wallet share customers. Also, historical peak sales could be helpful to identify the customer's business size.

The extracted features are:

1. *average_12_months*: customer average monthly spent in the last 12 months
2. *average_36_months*: customer average monthly spent in the last 36 months
3. *average_12_36_Ratio*: feature 1/feature 2
4. *top_1_month_spent*: the highest monthly spent in the last 36 months
5. *top_3_months_avg_spent*: the average of top 3 highest monthly spent in the last 36 months
6. *top_6_months_avg_spent*: the average of top 6 highest monthly spent in the last 36 months
7. *avg12_top_1_month_ratio*: feature 1/feature 4
8. *avg12_top_3_months_ratio*: feature 1/feature 5
9. *avg12_top_6_month_ratio*: feature 1/feature 6
10. *std_12_months*: the standard deviation of monthly spent
11. *spring_average*: the average monthly spent in Mar, Apr, May
12. *summer_average*: the average monthly spent in Jun, Jul, Aug
13. *autumn_average*: the average monthly spent in Sep, Oct, Nov
14. *winter_average*: the average monthly spent in Dec, Jan, Feb
15. *month_with_purchase_in_12_months*: no. of months with at least one purchase in the last 12 months
16. *month_with_purchase_in_36_months*: no. of months with at least one purchase in the last 36 months

The final column of the matrix was *wallet_share* representing the response value.

The first fourteen features (numbered from 1 to 14) come in two flavours: one that provides dollar values (and we use *dollar_* to precede the name of the feature), and the second one that provides the number of different product purchased (and we use *products_* to precede the name of the feature). For example, the original feature *average_12_months* is replaced by two features: *dollar_average_12_months* and *products_average_12_months*. So in total, we have 30 features columns (28 features generated from the first fourteen plus features 15 and 16) and *wallet_share* is the response column.

3.3   Estimation Model and Synthetic Data

The model is build using the Random Forest (RF) algorithm [9], with the number of trees set to 50. Training is done by 75%–25% random split and 10-fold cross validation. RMSE is used as the evaluation metric.

In addition, as the raw data is skewed, there are only a few cases for training sample of very low wallet share. This is very common in all real-world cases, since customers may not want to put low numbers on the survey. Thus, we have created synthetic data by introducing new empty entries and then set every column to a very low value. In addition to that, we also duplicated low wallet share samples and applied a small random variance to the feature values to create slightly different ones. Furthermore, there are many unlabeled customers in the transaction data that we could confidently assign a very low wallet share score without much analysis. For example, it is possible to assign a wallet share score of 0% to all customers that spent less than $100 during the last 12 months.

4   Results

In this section, the results of experiments are presented. First, Sect. 4.1 shows the importance of features for predicting the share of wallet. In Sect. 4.2, the accuracy of the developed method is compared with other existing methods. The impact of different training datasets and the additional synthetic data are investigated in Sects. 4.3 and 4.4, respectively.

4.1   Feature Selection

We have deployed the Boruta algorithm to select the most important features [8]. By default, Boruta runs Random Forest internally, testing each original feature against randomly generated features to check whether the original features can improve the prediction accuracy. Among the randomly generated features, the best performing one has been named ShadowMax and the mean performing random feature is called ShadowMean. Then, each original feature is compared with the randomly generated features. A feature that contributes positively should perform better than the best random feature (ShadowMax).

Results are shown in Fig. 2 for the paint dataset. Similarly, Fig. 3 shows the result for the air conditioner dataset. The randomly generated Shadow features are coloured as blue. We have validated the results with domain experts to assure the soundness of our approach.

Based on the comparison with the Shadow features, the features shown in green are confirmed, the features in yellow are questionable, whereas the algorithm suggests rejecting the features in red.

For the paint dataset, the *month_with_purchase_in12_months* is the best-performing feature, as expected by domain experts. The feature *dollar_std_12_months* is not performing as expected: it outperforms ShadowMax only by a small margin. After some discussions with domain experts, we concluded that this might indicate that the customer purchases do have a natural

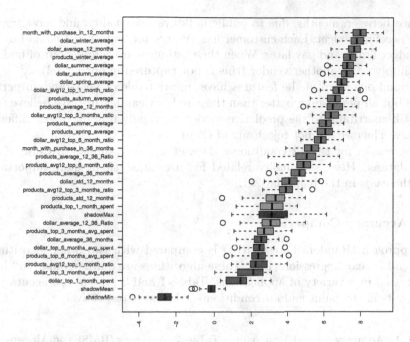

Fig. 2. Feature importance for the paint dataset. Shadow features have corresponding blue bars (Colors figure online)

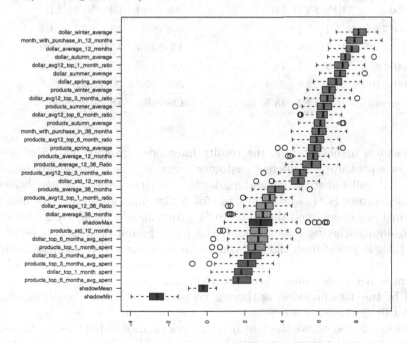

Fig. 3. Feature importance for air conditioner dataset. Shadow features have corresponding blue bars (Colors figure online)

variance between months, due to public holidays, seasonality, and most importantly account budget. Each customer has an account budget and could buy (to the budget limit) and pay later. When the customers are running out of budget, they simply go to another vendor (this is not captured in our model).

Around one-third of the features have questionable or near-random performance but all performing better than the ShadowMean feature. We believe they may still contribute to the predictive model, especially if applied to a different industry. Thus we did not reject any of them.

The results for the air conditioner dataset are very similar to those of the paint dataset. Here, seasonality related features show slight more importance than the cases in the paint dataset.

4.2   Accuracy Comparison

Our approach (Random Forest (RF)) is compared with two other algorithms: SVM and Linear Regression (LR). These algorithms are quite popular and have been tested in a variety of applications. Tables 1 and 2 present the accuracy of the results in the paint and air conditioning datasets, respectively.

Table 1. Accuracy (RMSE) on paint dataset

Paint	RF	SVM	LR
$0 < 15\%$	5.1	6.2	2.5
$15 < 50\%$	16.3	16.8	18.4
$50 < 80\%$	17.2	17.4	18.1
$80 < 100\%$	21.6	22.8	22.3
Overall	16.4	17.2	18.5

Table 2. Accuracy (RMSE) on Air conditioner dataset

Air cond.	RF	SVM	LR
$0 < 15\%$	3.2	5.1	4.4
$15 < 50\%$	17.9	19.3	18.9
$50 < 80\%$	17.8	20.1	21.1
$80 < 100\%$	20.3	19.9	20.5
Overall	18.7	19.2	20.5

As shown in both tables, the results have been grouped into 5 segments, each row representing a different customer segment. The segments are based on customers' wallet share value, and marketing activities usually target those segments separately. For example, the 0 to 15% wallet share group consists mainly of occasional customers, and the 80 to 100% group should be all loyal customers. Due to data gathering imperfections, the lower bound of RMSE is estimated to be 10%. It stems from the fact that some customers were sampled multiple times during the data gathering process and some sales representatives assigned different wallet share values when surveyed at different times. This has been confirmed by the data provider, as there is no perfect data, and the system should not over-fit the training set.

Random Forest shows the best overall performance in both test sets, and in nearly all the wallet share segments. However, it still scores an unimpressive 5.1

and 3.2 RMSE in the lowest wallet share group and 21.6 and 20.3 RMSE in the highest group. A limited number of training samples in those two groups are the main reason for such results.

SVM holds the middle position. In all but the 80% to 100% group in the air conditioner group, SVM has a lesser accuracy than Random Forest, but it seems to have less deviation between other groups.

Linear Regression shows the worst overall result, but it performs well in the lowest group (0 to 15%). This is probably a result of the use of synthetic data in the lowest group. But in all other groups, the performance of Linear Regression is inferior to the other two algorithms.

Additionally, the two businesses which provided the data have tested the model in real life. In both cases, this testing (evaluation) has been conducted for more than a year now, and the feedback has constantly been very positive. The estimation accuracy is in-line with the results shown in the tables.

Meanwhile, the data providers also keep collecting customers surveys and self-stated wallet shares are one of the focusing points. We found that the wallet share estimation system can also be used to identify inaccurate records in the survey results if the self-stated wallet share in the survey differs too much with the estimation. One of the examples of incorrect survey results could be: reporting 100% wallet share by a customer who has spent only 10 dollars with the business.

However, the users of our model have occasionally detected cases where the system appears to assign a low wallet share score to some of the known loyal customers who should have rather a high value of the wallet share. A detailed analysis of those individual cases often revealed that the customer in question has spent all his/her budget and stopped buying temporally. Due to the ad-hoc nature of such cases, we can consider them as outliers.

4.3   Training Size vs. Performance

Many businesses have overlooked the potential to apply data science to their operations. They may have heard the term big data and are afraid that they have not accumulated enough. Furthermore, it is unknown whether the inclusion of all historical data or all available data improves the accuracy of the results. Thus, we measured the accuracy of the results based on the varying size of the training data.

As shown in Fig. 4, the performance gradually reaches a plateau as the size of the training approaches 80. As we mentioned earlier, the model has been tested in real life. This indicates that it is possible to perform wallet share prediction with good accuracy even with a relatively small dataset.

Similar to the first dataset, the second dataset shows a gradually reducing error rate when the training size approaches 80. Figure 5 summarises this test.

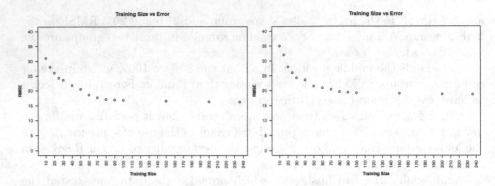

Fig. 4. Paint training size vs error

Fig. 5. Air conditioner training size vs error

4.4   Effect of Synthetic Data

As mentioned in Sect. 3.3, skewed data represent a common issue in real-world problems. In our case, we had very limited samples from the lowest group (wallet share between 0 and 15%), Table 3 presents the estimation accuracy using only the original data (paint dataset). Table 4 presents the estimation accuracy result in the air conditioner dataset using only the original data.

Table 3. Accuracy (RMSE) on paint dataset without synthetic data

Paint.	RF	SVM	LR
0 < 15%	9.5	8.7	6.4
15 < 50%	16.5	16.6	17.4
50 < 80%	16.9	19.2	18.3
80 < 100%	21.7	21.5	19.3
Overall	17.1	18.2	18.1

Table 4. Accuracy (RMSE) on air conditioner dataset without synthetic data

Air cond.	RF	SVM	LR
0 < 15%	7.4	6.4	6.2
15 < 50%	17.7	18.6	18.4
50 < 80%	17.8	19.5	20.4
80 < 100%	20.1	20.7	20.2
Overall	18.8	19.1	19.9

Table 5 presents a comparison between cases with and without the synthetic dataset on both original datasets (paint and air conditioner). Clearly, the performance improvement is significant, especially on the lowest wallet share group.

Table 5. Comparison: with/without synthetic data

	Paint		Air conditioner	
	With	Without	With	Without
0 < 15%	5.1	9.5	3.2	7.4
Overall	16.4	17.1	18.7	18.8

5 Conclusions

This paper presents a novel approach for the analysis and estimation of customer wallet share for manufacturers. Two major manufacturers which collect transaction data related only to their own products are investigated. The approach accurately predicts wallet shares scores based on transaction data only and does not rely on any additional survey data, as it is normally used in other approaches.

The proposed approach is evaluated using two real-world datasets. The first dataset consisted of the transactions data from one of the largest manufacturers of paint products in Australia. The other dataset came from a major Australian manufacturer and distributor of air conditioning products. Furthermore, the analysis of the most important features for wallet share estimation is provided. These findings can be helpful for similar problems as well. Additionally, it is shown that the proposed model can work with a limited training input and a data augmentation approach is presented to address the data skew issue. To the best of our knowledge, no existing research has investigated a similar problem with the same data limitations.

The contributions of this paper can be summarised as follows:

1. To the best of our knowledge, there is no similar research to estimate customer wallet share for manufacturers based on transaction data only, as described in the scenario. We developed a Random Forest predictive model and extended the existing features of transactions by creating new features.
2. We analysed and selected the most important features and provided a simplified and scalable model, which then was used to analyse a large customer base. This assisted in accurately estimating the share of wallet for customers for a manufacturer's product.
3. We showed that it was possible to build an estimation model with a small training set. Furthermore, we demonstrated the application of synthetic data to address the data skew problem.

This paper aims at encouraging companies to apply modern data science techniques in approaching their business problems and to start collecting more surveys to begin the process. We are currently experimenting with many other businesses to check the model's generalisation capability. In future research, we plan to augment transaction data with survey results for cross-referencing. The additional benefit of this step would be a possible identification of inaccurate survey entries. Moreover, using this data, a business could aim to outperform the original survey in wallet share estimation accuracy.

References

1. Baumann, C., Burton, S., Elliott, G.: Determinants of customer loyalty and share of wallet in retail banking. J. Financ. Serv. Mark. **9**(3), 231–248 (2005)
2. Chen, Y., Steckel, J.H.: Modeling credit card share of wallet: solving the incomplete information problem. J. Mark. Res. **49**(5), 655–669 (2012)
3. Çifci, S., Ekinci, Y., Whyatt, G., Japutra, A., Molinillo, S., Siala, H.: A cross validation of consumer-based brand equity models: driving customer equity in retail brands. J. Bus. Res. **69**(9), 3740–3747 (2016)
4. Glady, N., Croux, C.: Predicting customer wallet without survey data. J. Serv. Res. **11**(3), 219–231 (2009)
5. Keiningham, T.L., Cooil, B., Aksoy, L., Andreassen, T.W., Weiner, J.: The value of different customer satisfaction and loyalty metrics in predicting customer retention, recommendation, and share-of-wallet. Manag. Serv. Qual. Int. J. **17**(4), 361–384 (2007)
6. Keiningham, T.L., Perkins-Munn, T., Evans, H.: The impact of customer satisfaction on share-of-wallet in a business-to-business environment. J. Serv. Res. **6**(1), 37–50 (2003)
7. Keiningham, T.L., et al.: Perceptions are relative: an examination of the relationship between relative satisfaction metrics and share of wallet. J. Serv. Manag. **26**(1), 2–43 (2015)
8. Kursa, M.B., Rudnicki, W.R., et al.: Feature selection with the boruta package. J. Stat. Softw. **36**(11), 1–13 (2010)
9. Liaw, A., Wiener, M., et al.: Classification and regression by randomforest. R News **2**(3), 18–22 (2002)
10. Mägi, A.W.: Share of wallet in retailing: the effects of customer satisfaction, loyalty cards and shopper characteristics. J. Retail. **79**(2), 97–106 (2003)
11. Merugu, S., Rosset, S., Perlich, C.: A new multi-view regression approach with an application to customer wallet estimation. In: Proceedings of the 12th ACM SIGKDD International Conference on Knowledge Discovery and Data Mining, pp. 656–661. ACM (2006)
12. Rosset, S., Perlich, C., Zadrozny, B., Merugu, S., Weiss, S., Lawrence, R.: Wallet estimation models. In: International Workshop on Customer Relationship Management: Data Mining Meets Marketing (2005)
13. Subramaniam, L.V., Faruquie, T.A., Ikbal, S., Godbole, S., Mohania, M.K.: Business intelligence from voice of customer. In: 2009 IEEE 25th International Conference on Data Engineering, pp. 1391–1402. IEEE (2009)
14. Weiss, S.M., Indurkhya, N.: Estimating sales opportunity using similarity-based methods. In: Daelemans, W., Goethals, B., Morik, K. (eds.) ECML PKDD 2008. LNCS (LNAI), vol. 5212, pp. 582–596. Springer, Heidelberg (2008). https://doi.org/10.1007/978-3-540-87481-2_38

Curtailing the Tax Leakages by Nabbing Return Defaulters in Taxation System

Priya Mehta[1], Jithin Mathews[1], Sandeep Kumar[2], K. Suryamukhi[1], and Ch Sobhan Babu[1(✉)]

[1] Indian Institute of Technology Hyderabad, Sangareddy, India
{cs15resch11007,cs15resch11004,cs17mtech01002,sobhan}@iith.ac.in
[2] Plianto Technologies, Sangareddy, India
cs15mtech11017@iith.ac.in

Abstract. Tax evasion is an illegal activity where a taxpayer avoids paying his/her tax liability. Any taxpayer has to file their tax return statements periodically at regular intervals. Avoiding to file or delaying the filing of the tax return statement is one among the most basic methods of tax evasion. The taxpayers who are not filing returns or delaying the filing of returns are called return defaulters. Financial loss to the Government due to avoiding to file or delayed filing of returns varies between taxpayers. While designing any statistical model to predict return defaulters, we have to take into account the real financial loss associated with the misclassification. In this paper, we constructed an example dependent cost - sensitive logistic regression model that predicts whether a taxpayer is a potential return defaulter for the upcoming tax-filing period. While designing the model, we studied the effect of business interactions among the taxpayers on return filing behavior. We developed this model for the commercial taxes department, Government of Telangana, India. Applying our method to tax data, we show significant cost saving.

Keywords: Goods and services tax · Tax avoidance · Tax evasion · Logistic regression · Example dependent cost-sensitive logistic regression · Benford's analysis · Social network analysis · Trust rank

1 Introduction

Taxation systems are of two types: direct taxation system and indirect taxation system. In direct taxation system the taxpayer directly pays the tax to the Government (*e.g.*, Income tax) and in indirect taxation system tax is paid through a third party (*e.g.*, Commercial tax). GST (Goods and Services Tax) is an example of the indirect taxation system. In this paper, we deal with GST (Goods and Services Tax) [1], which is an indirect taxation system followed in India from July 2017.

T. D. Le et al. (Eds.): AusDM 2019, CCIS 1127, pp. 183–195, 2019.
https://doi.org/10.1007/978-981-15-1699-3_15

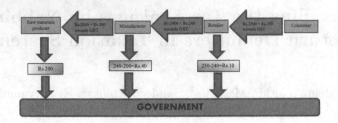

Fig. 1. Flow of tax in GST

1.1 How GST Works

In Fig. 1, we have taken an ornament manufacturer as an example and 10% as the GST rate. Note that, throughout the paper, we represent currencies in Indian currency "*Rupees*", denoted as "Rs.". The manufacturer purchases raw material of worth Rs. 2000 and pays GST of Rs. 200 (10% of 2000). Suppose the value addition happened in the manufacturing process to convert it into an ornament is Rs. 400. Hence, the value of the ornament is Rs. 2400. Now, the total tax on the sales of this ornament to the retailer is Rs. 240 (10% of 2400). By setting off the tax which he had already paid at the time of purchasing the raw material, the manufacturer is liable to pay GST of Rs. 40 (tax collected - tax already paid), *i.e.*, Rs. 40 (240–200). The retailer adds his margin of Rs. 100 making the total value to Rs. 2500 and sells it to the end customer by collecting Rs. 2500 towards the ornament and Rs. 250 towards tax. By setting off the tax he had already paid at the time purchasing ornament from the manufacturer, the retailer is liable to pay GST of Rs. 10 (tax collected - tax already paid), *i.e.*, Rs. 10 (250–240). Finally, GST received by the Government is Rs. 250, which is completely borne by the end customer.

1.2 Motivation for This Work

In GST system, taxpayers have to file their tax returns once in every month. By not filing the returns or delaying the filing of returns, the taxpayer benefits in three ways. First, they get sufficient time to manipulate their books; Secondly, the penalty imposed by the Government for late filing is negligible compared to the interest rates in the open market, and finally, possessing liquid cash is always advantageous to businesses like real estate. The motivation for this work is to construct a classification model to predict the potential return defaulters and following them to file returns by sending emails, SMS and physically visiting their business premises. We are working with the Government of Telangana, India, and analyzing their data sets and developing models to increase the compliance level of return filing. We used techniques from social network analysis to create independent variables that capture the interaction of a taxpayer with other taxpayers. We also used the taxpayer's characteristics, like, average tax per month, total sales amount, etc., in creating the independent variables required to design the model.

It is a very common practice to use standard cost-insensitive binary classification algorithms like logistic regression and decision trees to design classifiers. However, in practice, the cost associated with classifying genuine taxpayer as return defaulters is quite different from classifying a return defaulter as a genuine taxpayer. Again cost associated with misclassifying a return defaulter as a genuine taxpayer varies from taxpayer to taxpayer based on their turnover. To capture this scenario, we designed an *example dependent cost-sensitive* logistic regression model [2–4]. We built an example-dependent cost-sensitive logistic regression model that predicts with high accuracy whether a taxpayer is going to file their return in the upcoming month. The taxation authorities can use this information to take proactive measures like sending SMS and emails to potential defaulters to increase the compliance levels.

Indirect taxation system is quite common in several nations. The approach we followed in this paper can be generalized to indirect taxation systems followed in other nations.

Our paper is structured as follows. In Sect. 2, we brief on existing works that are related to ours. Section 3 discusses the methodology involved in the design of this predictive model. Section 4 discusses the model along with model performance. Finally, the concluding remarks are provided in Sect. 5.

2   Related Work

In [5], Lismont et al. designed analytical models to identify the tax evasion. They constructed a network of firms connected through shared board membership and showed that including network effects significantly improves the predictive ability of tax avoidance models. In [6], Bianchi et al. constructed a network of auditors and used network centrality measures to show that firms engaging better-connected auditors exhibit comparatively lower effective tax rates. In [7], Vlasselaer et al. worked on social security fraud. They aimed to identify fraudulent business entities by propagating a time-dependent exposure score for each business entity based on its relationships to known fraud business entities in the social network. In [8], Sahin and Duman designed classification models using logistic regression and Artificial Neural Networks for credit card fraud detection. This study is one of the firsts to compare the performance of ANN and LR methods in credit card fraud detection with a real data set. In [9], Wilson worked on insurance fraud in the auto industry. He looked at various situations and various tactics that are used by insured people to defraud insurance companies. They used logistic regression to detect fraudulent claims. In [10], van der Meulen et al. gave a detailed explanation of a project executed by them towards improving participation in a rehabilitation program by the patients of the Virga Jesse Hospital. They designed logistic regression model and showed that the major influential factor is the distance from the patient's home. The class-imbalanced data sets occur in many real-world applications, where the class distributions of data are highly imbalanced. In [11], Ling and Sheng showed that cost-sensitive learning is a common approach to solve this problem. In [2], Bahnsen et al. proposed an example dependent cost matrix for credit scoring. They proposed a cost function that introduces the example dependent costs into logistic regression.

In this paper, we constructed an example dependent cost-sensitive logistic regression model that predicts whether a taxpayer is a potential return defaulter for the upcoming tax-filing period. While designing the model, we constructed a network of tax payers and studied the effect of business interactions among the taxpayers on return filing behavior.

3   Methodology and Feature Extraction

Our objective is to design an *example dependent cost-sensitive* logistic regression model to predict whether a given taxpayer (business entity) will file GST return or not in the coming month. The reason for choosing an *example dependent cost-sensitive* logistic regression is misclassification costs, in particular, false-negative costs, which are not the same for every taxpayer [2]. We designed this model based on the taxpayers past returns filing behavior, the volume of sales and purchases, the value of business interactions among taxpayers and *Mean absolute deviation* (*MAD*) value of the first digit Benford's analysis on the sales transactions of taxpayers. This model is developed for the commercial taxes department of Telangana, India. For the same, we used the data set provided by them that contains mainly two tables.

In Subsect. 3.1, *example dependent cost-sensitive* logistic regression is explained. In Subsect. 3.2, we explain a statistical fraud detection technique called Benford's Law, which we use as one of the technique for feature extraction. In Subsect. 3.3, we give the description of the data used towards building an example dependent cost-sensitive logistic regression model. In Subsect. 3.4, we will explain the construction of a social network of taxpayers to enable the computation of value of business interactions of a taxpayer with other taxpayers, which is another significant independent variable in the model. In Subsect. 3.5, we explain the set of independent variables (features) used to design the model and methodology to derive them.

3.1   Example Dependent Cost-Sensitive Logistic Regression

Logistic regression is a classification model that estimates the probability of the positive class. The estimated probability of the taxpayer X_i belonging to the positive class is given by Eq. 1

$$h_\theta(X_i) = 1/(1 + e^{-(\theta_0 + \sum_{j=1}^{m} \theta_j x_i^j)})$$

(1)

where m is the number of independent variables, θ is the coefficients vector, and x_i^j is the value of the jth independent variable of the taxpayer X_i. The coefficients vector θ is estimated to minimize the cost function in Eq. 2

$$\sum_{i=1}^{n} -y_i log(h_\theta(X_i)) - (1 - y_i) log(1 - h_\theta(X_i)) \qquad (2)$$

where y_i is the actual class of the taxpayer X_i and n is the number of tax-payers. This cost function assigns the same weight to different errors, both false-positives, and false-negatives. It assumes a constant cost of misclassification errors.

Predicting return defaulter is an example dependent cost-sensitive model by nature, where the costs due to misclassification vary between taxpayers. In this scenario, we select the coefficients vector θ to minimize the cost function in Eq. 3

$$\sum_{i=1}^{n} y_i(h_\theta(X_i) * CTP_i + (1 - h_\theta(X_i)) * CFN_i) +$$

$$(1 - y_i)((1 - h_\theta(X_i)) * CTN_i + h_\theta(X_i) * CFP_i) \quad (3)$$

Here $CTP_i, CFN_i, CTN_i, CFP_i$ are true-positive, false-negative, true-negative, and false-positive costs of the taxpayer X_i, respectively. In the cost function 3, when the actual class of X_i is *class one (positive class i.e. $y_i = 1$)* then the contribution by X_i towards the total cost is given by $h_\theta(X_i) * CTP_i + (1 - h_\theta(X_i)) * CFN_i$. This value is the sum of two quantities, one is the predicted probability of X_i belonging to class one (positive class) multiplied by the true-positive cost of X_i, and the other is the predicted probability of X_i belonging to class zero (negative class) multiplied by the false-negative cost of X_i.

3.2   Benford's Analysis

Benford's law is a mathematical method for identifying fraud [12-14] in naturally-occurring numbers, which are neither highly constraint nor purely random. This law states that the percentage of numbers starting with the digit $k \in \{1, 2, ..., 9\}$ follows the formula $log_{10}(1 + 1/k)$.

Mean absolute deviation(MAD) is a statistical technique to verify if the distribution of data's first digits follows the expected probability distribution given by Benford's law. Data is segregated into nine bins depending on the first digit. $MAD = \sum_{j=1}^{9}(AP_j - EP_j)/9$, where AP_j denotes the observed portion of the jth bin, and EP_j denotes the expected portion of the jth bin.

3.3   Description of Data Set

We used two types of data sets to design the model. One is GSTR-1 data and other is the monthly GST returns data.

GSTR-1 Data: GSTR-1 is a monthly financial statement that should be submitted by every taxpayer. This statement contains details of all outward supplies, i.e., sales done during the corresponding month of this statement. The following table contains some fields of this data set. Every row in Table 1 corresponds to one sales transaction. The data set we have taken contains several millions of such rows. The actual statement contains more information, such as the rate of tax, the quantity of goods sold, *etc.*

Table 1. GSTR-1 data

S. No	Month	Seller	Buyer	Invoice number	Amount(Rs)
1	Jan 2018	A	B	XY123	12000
2	Feb 2018	B	D	ZU342	18000
3	Jan 2018	B	C	UX5434	14000
4	July 2018	C	D	YS8779	15000
5	Mar 2018	D	A	ZX7744	12000

Monthly GST Returns Data: Table 2 contains a few fields of GST returns data. Each row in this table corresponds to a monthly return by a taxpayer. *ITC (Input tax credit)* is the amount of tax the taxpayer paid during purchases of services and goods. The *output tax* is the amount of tax the taxpayer collected during the sales of services and goods. The taxpayer has to pay to the Government the gap between the *ITC* and *output tax*, i.e. output tax - ITC. The actual database consists of much more information, like, tax payment method, return filing data, international exports, exempted sales, and sales on RCM (reverse charge mechanism).

Table 2. GST Returns data

S.No	Firm	Month	Purchases	Sales	ITC	Output tax
1	A	Feb-18	180000	220000	20000	26000
2	D	Sep-18	200000	280000	5000	9000
3	E	Oct-17	400000	480000	40000	48000

3.4   Creation of Network of Taxpayers

One of the independent variables in our model is the amount of business interaction a taxpayer has with other taxpayers. To compute this independent variable, we created a weighted, directed graph (social network). Each vertex (node) in this directed graph corresponds to a taxpayer. The weight assigned to a vertex is the average tax paid per month [*ATPM*] by the corresponding taxpayer during the period July 2017 to April 2019. Vertex weights are normalized using min-max normalization. We used the monthly GST Return Data explained in

Table 2 to compute the vertex weights. We colored each vertex as *BLACK* or *YELLOW* or *WHITE*. Vertex color *Yellow* means the corresponding taxpayer filed at least three-fourths of GST returns ($\frac{3}{4} * 22 = 17$ returns during July 2017 to April 2019) in-time, vertex color *Black* means that the corresponding taxpayer has filed at most one-fourth of GST returns ($\frac{1}{4} * 22 = 5$ returns during July 2017 to April 2019) in-time and the rest of the vertices are colored as *White*. These ratios (three-fourths and one-fourth) are chosen empirically based on model performance. This coloring can be done using the monthly GST Return data explained in Table 2. We placed a weighted directed edge from taxpayer a to taxpayer b, where edge weight is the amount of sales done by taxpayer a to taxpayer b during the period July 2017 to April 2019. Then the min-max normalization of edge weights is performed. For the same, we used the sales data explained in Table 1. This graph will capture the scale of interaction and(or) the exchange of money between taxpayers.

3.5   Feature Extraction

Figure 2 explains the variables (features) used in the model. Let b be the vertex for which we are constructing the features mentioned in Table 2. Below is the detailed explanation for each variable (feature).

S No	Feature	Min	Max	Mean	Varaince
1	2	3	4	5	6
1	Filed	0	1	-	-
2	Ratio	0.001	201.000	0.967	8.400
3	ATPM	0.000	1.000	0.256	0.135
4	Not Filed Count	0	18	6.670	42.330
5	Total Purchase Amount	0.000	359.800	0.290	5.520
6	MAD Value	0.003	0.022	0.010	0.0009
7	Division Name	-	-	-	-

Fig. 2. Variables

Ratio: This is the variable extracted from the graph defined in Subsect. 3.4. This graph captures the degree of interaction and(or) the money exchange between taxpayer b and other taxpayers. This variable captures the influence of other taxpayers on b. If b has close ties with taxpayers who are not filing GST returns in-time, then, they would influence b not to file GST returns and viceverse [5]. Let B be the set of all BLACK colored vertices in the graph constructed in Subsect. 3.4 and Y be the set of all vertices, which are colored YELLOW.

- $b_{11} = \sum_{v \in B} \frac{w(v) * w(vb)}{w(v) + w(vb)}$, where $w(v)$ is the weight of vertex v and $w(vb)$ is the weight of directed edge vb
- $b_{12} = \sum_{v \in B} \frac{w(v) * w(bv)}{w(v) + w(bv)}$.
- $b_{21} = \sum_{v \in Y} \frac{w(v) * w(vb)}{w(v) + w(vb)}$, where $w(v)$ is the weight of vertex v and $w(vb)$ is the weight of directed edge vb
- $b_{22} = \sum_{v \in Y} \frac{w(v) * w(bv)}{w(v) + w(bv)}$.

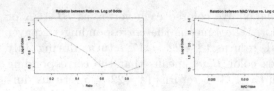 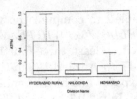

Fig. 3. Ratio vs Log of odds **Fig. 4.** MAD vs Log of odds **Fig. 5.** Division vs ATPM

The variable *Ratio* for vertex b is $\frac{b11+b12}{b21+b22}$. Higher the value of *Ratio* means b is doing huge business transactions with return defaulters who can influence b to not to file tax return in-time. The relation between *Ratio* variable and *Log of Odds* of dependent variable *Filed* is given in Fig. 3. This is an asymptotic relation. Based on this observation, we used the log of *Ratio* as an independent variable in model building. Note that we did not use the vertices colored WHITE in computing this feature, as their return filing behavior sits between non-filing behavior of BLACK vertices and in-time filing behavior of YELLOW vertices. Influence of vertices colored WHITE on vertex b can be ignored.

Filed: This is the *dependent variable* in the model with a binary outcome. This variable gives the GST return filing status (filed in-time/not filed in-time) of the taxpayer b for the month of May 2019. *Zero* denotes filed in-time (negative class) and, *one* denotes not filed in-time (positive class). Note that in the data set there are 25% class one records and rest are class zero records. Note that 25% of taxpayers did not file returns before the due date in May-2019, which resulted in a loss of few hundreds of millions of rupees to the Government.

Not Filed Count: This is the number of GST returns not filed in-time before the due date of the corresponding month by b from July-2017 to April-2019.

Division-Name: Telangana state is segregated into 12 geographic divisions for simplifying the administration works. This independent variable gives the division name of b, where b is located.

ATPM: This is the average tax per month paid by b. We included square, cube, and the square root of *ATPM* in the model as the relation between *ATPM* and *Log of Odds* of the dependent variable is a polynomial.

Total Purchase Amount: It is the total amount of purchases by b from July 2017 to April 2019. The relation between *Total Purchase Value* and *Log of Odds* of the dependent variable is a linear relation.

MAD Value: It is the Mean absolute deviation value of the first digit benford's law on sales transactions of b. From Fig. 4, One can observe that the relation between *MAD Value* and *Log of Odds* of dependent variable *Filed* is a non-linear relation.

4   Experimental Results

4.1   Software Used

We used the open-source statistical software R to build the model [15, 16]. Because of its flexibility, extensibility, and no cost, it has gained a lot of popularity in the industry as well as in academic environments.

4.2   Costs Matrix

Table 3 gives different miss-classification costs of a given taxpayer.

- **True-negative cost** is zero. Classifying an in-time return filer (actual class zero) as an in-time return filer (predicted class zero) will not cost us anything.
- **True-positive cost** is expenses towards sending SMS, calling the taxpayer, and associated manpower cost. This cost is the same for all taxpayers whose actual class is one and predicted class is one. This cost is Rs. 150.
- **False-positive cost** is expenses towards sending SMS, calling the taxpayer, and associated manpower cost. This is the same for all taxpayers whose actual class is zero, and predicted class is one. This cost is Rs. 150.
- **False-negative cost** depends on the ATPM of the taxpayer and the expected number of days of delay in filing return by this taxpayer. This is given by $\frac{ATPM*expected\ number\ of\ days\ of\ delay*18}{36500} * 3 + 100$.
  Here $\frac{ATPM*expected\ number\ of\ days\ of\ delay*18}{36500}$ is the loss of interest due to late filing of return, where interest rate is 18%. This cost varies from taxpayer to taxpayer as ATPM and expected number of days of delay in filing return vary from one taxpayer to the other. We multiplied the loss of interest by three times and added 100 to it, in order to minimize the chance of a defaulter becoming a chronic defaulter.

Table 3. Example dependent miss-classification costs matrix

	Actual class 1	Actual class 0
Predicted class 1	150	150
Predicted class 0	CFN	0

4.3   Model Parametric Coefficients

We computed θ to minimize the cost function in Eq. 4

$$\sum_{i=1}^{n} y_i(h_\theta(X_i)*150+(1-h_\theta(X_i))*CFN_i)+(1-y_i)((1-h_\theta(X_i))*0+h_\theta(X_i)*150)$$

$$(4)$$

In this equation, CFN_i is the false-negative cost of taxpayer X_i and y_i is the actual class of X_i.

Figure 6 gives the parametric coefficients.

S No	Coefficients	Estimate
1	2	3
1	Intercept	-1.091
2	Log(Ratio)	15.323
3	MAD value	-16.692
4	Square(MAD value)	-1.793
5	Cube(MAD value)	-4.449
6	ATPM	13.166
7	Log(ATPM)	1.221
8	Square(ATPM)	6.404
9	Cube(ATPM)	16.187
10	Sqrt(ATPM)	-12.866
11	Not Filed Count	-17.585
12	Total Purchase Amount	-28.341
13	ATPM * Hyderabad Rural Division	-5.037
14	ATPM * Nizamabad Division	46.054

Fig. 6. Parametric coefficients

From Fig. 5, it can be inferred that the distribution of *ATPM* is not the same in every division. Interaction variables are created by multiplying dummy variables corresponding to *Division Names* and *ATPM* to handle this. Even though there are twelve divisions in Telangana, the interaction between only two divisions and *ATPM* are statistically significant. Variables 13 and 14 in Fig. 6 are these two interaction variables.

4.4   Model Validation

Confusion and Cost Matrices: Figures 7 and 8 are the training and the testing confusion matrices at cut off equal to 0.6. The training accuracy of the model is 81.35%, and the testing accuracy is 81.37%.

Figures 9 and 10 are the training cost, and the testing cost matrices at the cut off equal to 0.6. These give the true-positive cost, false-negative cost, true-negative cost and false-positive cost of both the training set and testing set. The average training cost is 68.29 rupees, and the average testing cost is 68.72 rupees.

	Predicted 0	Predicted 1
Actual 0	62672	10331
Actual 1	8980	21612

	Predicted 0	Predicted 1
Actual 0	26780	4432
Actual 1	3835	9351

	Predicted 0	Predicted 1
Actual 0	0	1549650
Actual 1	2273832	3241800

	Predicted 0	Predicted 1
Actual 0	0	664800
Actual 1	983682	1402650

Fig. 7. Training confusion matrix   **Fig. 8.** Testing confusion matrix   **Fig. 9.** Training cost matrix   **Fig. 10.** Testing cost matrix

Training and Testing ROC Curves: Training and testing ROC curves are given in Fig. 11. AUC value of training ROC curve is 0.79 and AUC value of testing ROC curves is also 0.79. From these observations, One can conclude that the model is neither under fitting nor over-fitting. Concordance value on test data is 0.70.

Gain, Lift and Kolmogorov–Smirnov charts are given in Figs. 13, 12, 15, 14, 17 and 16.

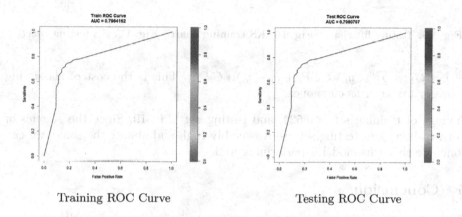

Training ROC Curve   Testing ROC Curve

Fig. 11. ROC curves

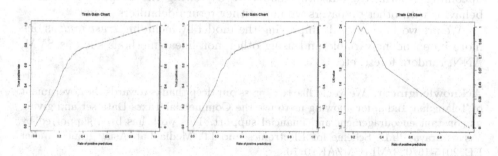

Fig. 12. Training gain chart   **Fig. 13.** Testing gain chart   **Fig. 14.** Training lift chart

Savings: Towards measuring the performance of an *example dependent cost-sensitive* logistic regression, we use the *savings* of the model. The *savings* of the model [2] is computed as : $\frac{min(C_0, C_1) - C_{opt}}{min(C_0, C_1)}$.

- Let $C_{opt} = \sum_{i=1}^{n} y_i(c_i * CTP_i + (1 - c_i) * CFN_i) + (1 - y_i)((1 - c_i) * CTN_i + c_i * CFP_i)$, where y_i is the actual class, and c_i is the predicted class of the ith taxpayer.
- Let $C_0 = \sum_{i=1}^{n} y_i * CFN_i + (1 - y_i) * CTN_i$. This is the cost of classifying every taxpayer as class zero.

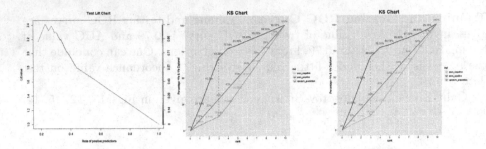

Fig. 15. Testing lift chart   **Fig. 16.** KS training chart   **Fig. 17.** KS testing chart

– Let $C_1 = \sum_{i=1}^{n} y_i * CTP_i + (1 - y_i) * CFP_i$. This is the cost of classifying every taxpayer as class one.

Savings of training set is 0.524, and testing set is 0.519. Since the *savings* of both training and testing set are reasonably high and almost the same, we can conclude that this model is performing well.

5 Conclusion

We built an *example dependent cost-sensitive* logistic regression model that predicts whether a given taxpayer is a plausible return defaulter or not for the upcoming month. We built the model by exploiting the taxpayer's business behavior with other taxpayers who are either return defaulters or not.

We are working towards improving the model by including *trust rank* as one more independent variable and using other non-linear methods such as SVM, ANN, random forest, etc.

Acknowledgment. We would like to express our deep thanks towards the government of Telangana, India, for allowing us to use the Commercial Taxes Data set and giving us constant encouragement and financial support. This work has been supported by Visvesvaraya Ph.D. Scheme for Electronics and IT, Media Lab Asia, grant number EE/2015-16/023/MLB/MZAK/0176.

References

1. Dani, S.: A research paper on an impact of goods and service tax(GST) on Indian economy. Bus. Econ. J. **7**, 264 (2016). ISSN: 2151–6219
2. Bahnsen, A.C., Aouada, D., Ottersten, B.: Example-dependent cost-sensitive logistic regression for credit scoring. In: 2014 13th International Conference on Machine Learning and Applications, pp. 263–269. IEEE, December 2014. ISBN: 978-1-4799-7415-3
3. Günnemann, N., Pfeffer, J.: Cost matters: a new example-dependent cost-sensitive logistic regression model. In: Kim, J., Shim, K., Cao, L., Lee, J.-G., Lin, X., Moon, Y.-S. (eds.) PAKDD 2017. LNCS (LNAI), vol. 10234, pp. 210–222. Springer, Cham (2017). https://doi.org/10.1007/978-3-319-57454-7_17

4. Scott, C.: Surrogate losses and regret bounds for cost-sensitive classification with example-dependent costs. In: ICML (2011)
5. Lismont, J., Cardinaels, E., Bruynseels, L., Groote, S.D., Lemahieu, W., Vanthienen, J.: Predicting tax avoidance by means of social network analytics. Decis. Support Syst. **108**, 13–24 (2018)
6. Bianchi, P., Falsetta, D., Minutti-Meza, M., Weisbrod, E. H.: Professional networks and client tax avoidance: evidence from the Italian statutory audit regime (2016). https://ssrn.com/abstract=2601570
7. Vlasselaer, V.V., Akoglu, L., Eliassi-Rad, T., Snoeck, M., Baesens, B.: Guilt-by-constellation: fraud detection by suspicious clique memberships. In: 48th Hawaii International Conference on System Sciences HICSS, pp. 918–927. IEEE, January 2015. ISBN: 978-1-4799-7367-5. https://doi.org/10.1109/HICSS.2015.114
8. Sahin, Y., Duman, E.: Detecting credit card fraud by ANN and logistic regression. In: 2011 International Symposium on Innovations in Intelligent Systems and Applications. IEEE, June 2011. ISBN: 978-1-61284-919-5
9. Wilson, J.H.: An analytical approach to detecting insurance fraud using logistic regression. J. Financ. Account. **1**, 1 (2009)
10. van der Meulen, F., Vermaat, T., Willems, P.: Case study: an application of logistic regression in a six sigma project in health care. Qual. Eng. **23**, 113–124 (2011)
11. Ling, C.X., Sheng, V.S.: Cost-sensitive learning and the class imbalance problem. In: Encyclopedia of Machine Learning. Springer, Heidelberg, January 2008
12. Nigrini, M.J., Mittermaier, L.J.: The use of Benford's law as an aid in analytical procedures. Audit. J. pract. Theory **41**, 52 (1997)
13. Asllani, A., Naco, M.: Using Benford's law for fraud detection in accounting practices. J. Soc. Sci. Stud. **1**, 129–143 (2014)
14. Durtschi, C., Hillison, W.: Pacini, C: The effective use of Benford's law to assist in detecting fraud in accounting data. J. Forensic Account. **5**(1), 17–34 (2004)
15. Ihaka, R., Gentleman, R.: R: a language for data analysis and graphics. J. Comput. Graph. Stat. **5**, 299–314 (1996)
16. Wood, S.N. (ed.): Generalized Additive Models: An Introduction With R. Chapman and Hall/CRC Press, Boca Raton (2006)

Network Path Estimation in Uncertain Data via Entity Resolution

Dean Philp[1], Naomi Chan[1], and Wolfgang Mayer[2]([⊠]) [iD]

[1] Defence Science Technology Group, Edinburgh, Australia
{dean.philp,naomi.chan}@dst.defence.gov.au
[2] University of South Australia, Adelaide, SA, Australia
wolfgang.mayer@unisa.edu.au

Abstract. Network Path Estimation is the problem of finding best paths among multiple potential routes between two devices, which is important to cyber situational awareness. In this context, information obtained from multiple sources and at different points in time must be integrated. However, duplicate representations of the same entities in different data sources must be identified and merged to accurately infer and rank network paths. We extend previous work on deterministic rule-based Entity Resolution with similarity flooding principles to obtain a probabilistic entity matching technique. Our approach outperforms the rule-based approach, allows for domain-specific ontologies to be incorporated, and accounts for provenance across data sources. Using the probabilistic resolutions, we rank network paths according to certainty of the resolutions, which improves network path estimation and contributes to cyber situational awareness.

Keywords: Entity Resolution · Path Estimation · Similarity Flooding · Contextualized Data

1 Introduction

Network paths refer to the traversed routes – hopping via intermediate devices— of network traffic between a given source and destination in computer networks. For example, using common Virtual Private Networking (VPN) software, home-based employee workstation may communicate with an employer server by routing across the Internet. Network paths traversing multiple unknown intermediate networks and devices present both a concern to cyber defensive groups and an opportunity for organised adversaries, as traffic interception and/or manipulation are potential threats. Cyber situational awareness for networks includes techniques for understanding network paths. However, this is a challenging problem. Without cooperation of intermediate network owners, network knowledge techniques have to rely on multiple disparate heterogeneous network data sources to infer potential traversed routes.

The information obtained from data sources must also be contextualized as it is valid only from the perspective of each measurement time, location and

© Springer Nature Singapore Pte Ltd. 2019
T. D. Le et al. (Eds.): AusDM 2019, CCIS 1127, pp. 196–207, 2019.
https://doi.org/10.1007/978-981-15-1699-3_16

other parameters. In the presence of uncooperative intermediate network owners (and/or adversarial entities such as hacker groups), we are confronted with
further data reliability issues and uncertainty of measurements.

Contextualized measurements can be accurately described using formal
knowledge representation techniques. GraphSource is a technique based on
named-graphs for capturing contextualized knowledge using RDF quads [12,13].
GraphSource defines explicit semantics of the fourth RDF element, enabling
information fusion and automated reasoning across contextualized knowledge.
However, duplicate entities represented multiple times in different GraphSources
may be frequent, especially when two GraphSources have related provenance, as
information is captured on repeatedly from the same network or otherwise overlapping measurement coverage over time.

Entity Resolution across GraphSources is vital for providing network analysts with decision support tools for network path recommendations [4]. However, given that information sources in this context suffer from incompleteness
and varying levels of reliability (quality), the task of inferring reliable link information can be difficult. Although rule-based linking of GraphSources based on
logical dependencies can accurately resolve duplicate entities if all sources provide reliable and consistent information, the presence of data quality issues [15]
renders the approach less effective. Instead, structural, syntactic and/or statistical relationships must be considered.

In this paper, we show that Similarity Flooding [8]—a probabilistic entity
resolution technique—can be applied across GraphSources to improve network
path estimation. We demonstrate an entity resolution method that is inspired
by similarity flooding and incorporates a rule-based domain-specific knowledge
into the matching process. Our method outperforms previous deterministic rule-
based entity resolution approach on imperfect data [4]. A salient feature of our
work is the use of provenance meta-data to inform the similarity calculation.
Moreover, we investigate the impact of different strategies for extracting matches
from inferred similarity scores.

We focus on the identification of duplicated representations of the same network elements among a set of data sources captured from network sensors over
time, assuming that ontological homogeneity has already been established among
the data sets. Specifically, we use RDF and the Computer Network and For-
warding Ontology (CNTFO) [13] for data representation throughout this paper.
CNTFO is used to filter unnecessary comparisons, seed the pairwise connectivity graph and to provide targeted thresholds for each entity type, akin to
blocking strategies in well-known entity resolution frameworks (see Sect. 3 for
more details). However, the presented approach is independent from the specific
ontology and can extend to other domains where different feature combinations,
weights, and thresholds may apply.

Using similarity measures we can rank network paths that contain duplicates
with highest certainty resolutions at the top. When making decisions about
preferred best network paths, this ranking provides insight into the inherent
uncertainty level in the underlying data (see Sect. 5).

2   Uncertainty in Network Data

The ISPNet dataset [14] is a semantically rich dataset of network control-plane data. In contrast to user-plane data, typically derived from high-volume network measurements, control-plane data is orders of magnitude smaller but more detailed. ISPNet contains four types of disparate data sources; (1) network configuration; (2) network paths; (3) network address requests; (4) routing protocols. Using network measurements from these data sources, ISPNet contains 43 GraphSource graphs. A PROVENANCE graph captures context, including time and location. Whilst ISPNet contains only 770 entities in total, it has semantic richness; there are around 2,500 statements about those entities, including around 870 relationships between entities.

Real-world network data is variable in reliability, consistency and often contains duplicates. Uncertainty arises from incomplete data, conflicting data, and different representations of the same network elements gathered from multiple network measurements taken in different contexts. For example, in Listing 1.1, duplicate entities on lines 2, 5 and 8 come from different contexts—different hosts and times (lines 12–17). Moreover, the fact on line 10 (net:hasMACAddress) is absent in the other collected graphs and hence cannot aid in entity resolution. These uncertainties motivate the use of probabilistic Entity Resolution to provide a measurement of similarity and support the ranking network paths.

Listing 1.1. ISPNet sample

```
1   ispnet:CORE {            //GraphSource 1: network configuration
2       core:C1-ADL-PC3 a net:Computer .
3       core:C1-ADL-PC3_eth0 net:ipv4 "10.10.0.67"^^net:ipv4Type . }
4   ispnet:TRACEROUTE6 {     //GraphSource 2: network paths
5       traceroute6:NE_C1-ADL-PC3 net:hasInterface traceroute6:I10.10.0.67 .
6       traceroute6:I10.10.0.67 net:ipv4 "10.10.0.67"^^net:ipv4Type . }
7   ispnet:ARPING1 {         //GraphSource 3: network address requests
8       arping1:NE_C1-ADL-PC3 net:hasInterface arping1:I10.10.0.67 .
9       arping1:I10.10.0.67 net:ipv4 "10.10.0.67"^^net:ipv4Type ;
10          net:hasMACAddress "00:00:00:aa:00:1f" ; }
11  ispnet:PROVENANCE {      //context information
12      ispnet:CORE net:importHost "CORE" .
13      ispnet:CORE net:importTime "2018-05-14T16:44:23"^^xsd:dateTime .
14      ispnet:TRACEROUTE6 net:importHost "C1-ADL-PC3" .
15      ispnet:TRACEROUTE6 net:importTime "2019-02-14T16:43:11"^^xsd:dateTime
            .
16      ispnet:ARPING1 net:importHost "C1-ADL-PC3" .
17      ispnet:ARPING1 net:importTime "2019-02-14T16:42:38"^^xsd:dateTime . }
```

3   Approach

This section describes how we adapt the Similarity Flooding algorithm in [8]; called *SimFlood*. The similarity flooding algorithm starts with an initial similarity guess for each pair of possible duplicates, and propagates this guess among a graph representing potential matches. The propagation is iterated until the similarity measures are stable.

Algorithm 1. SimFlood pseudocode

1: entityTypes ← [Router,Network,Interface];
2: comparisonAttr ← {Interface: ipv4,
                          Router: [hasInterface, ipv4, routerId],
                          Network: ipv4subnet,
                          LITERAL: None };   ▷ Domain Knowledge
3: baseGraph ← parseRDF(file);
4: listOfProvenances ← groupProvenances(baseGraph);
5: **for** graphList in listOfProvenances **do**
6:    **for** g1 in graphList **do**
7:       **for** g2 in graphList **do**
8:          pcGraph ← PairwiseConnectivityGraph(g1, g2, comparisonAttr);
9:          pcGraph.simFlood(15);
10:         **for** entity in entityTypes **do**
11:            threshold ← pcGraph.calculateRelativeThreshold(entity);
12:            pcGraph.writeResults(threshold, entity);
13:         **end for**
14:      **end for**
15:   **end for**
16: **end for**

Algorithm 1 provides pseudo-code of how the similarity flooding algorithm was adapted. Line 1 constrains the RDF entity types of potential duplicates; these types are extracted from CNTFO. We assume GraphSource datasets will contain many duplicated *Routers, Networks* and *Interfaces* CNTFO-typed instances from multiple heterogeneous data sources, where duplicates should be merged due to non-overlapping semantics arising from each data source. These CNTFO types could be input parameters to facilitate other domain-specific ontologies. These types are also used to constrain the possible comparisons. We do not compare duplicates of type *Router* with *Interface*, but will compare duplicates of type *Router* with *NetworkElement* – the former is a subclass of the latter. Line 2 constrains the comparison attributes for each CNTFO type. This is also a domain-specific input parameter. The comparison attributes are used to construct the Pairwise Connectivity Graph (see Fig. 1). For each type, different comparison attributes contribute to the initial seeding of similarity. The intuition here is that if a feature such as a Router IPv4 address matches, then these are more likely duplicates. Line 3 reads the input RDF file, then line 4 groups named graph by provenance to start the loop on line 5. The intuition here is that we do not want to consider duplicates across data sources that have completely different measurement time and/or location. For example, data from last year is probably irrelevant this year in the context of a business-network that has since changed. Likewise, data from Queensland is probably irrelevant in Victoria unless we have a national network across large geographical boundaries. Regardless, the context information—data provenance—can be used to constrain sensible comparisons in a domain-specific way. On line 8, we create the Pairwise Connectivity Graph then run the main Similarity Flooding

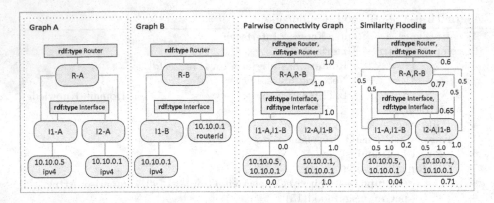

Fig. 1. Creating the similarity flooding graph

algorithm on line 9. Figure 1 provides an example. The first two graphs are created using aforementioned domain-specific knowledge from CNTFO; entityTypes and comparisonAttr are elements of our ontology that encode which duplicates we want to resolve and attributes that are comparable (see lines 1–2 in Algorithm 1). Note that included in both graphs are two rdf:type nodes—*rdf:type Router* and *rdf:type Interface*. In Graph A we have a Router *R-A* with two Interfaces *I1-A* and *I1-B*, each with an ipv4 address. In Graph B we a have Router *R-B* with one Interface (and associated ipv4 address), and a routerId. In the Pairwise Connectivity Graph we combine the same type comparison attributes (in this case only Interface and not routerId), resulting in initial seeds (0.0 or 1.0) by comparing comparison attribute values (e.g. 10.10.0.1 equals 10.10.0.1). Note that the comparison of attribute values is also domain-specific. In our example, the domain-specific knowledge is that if two Routers are discovered with comparable provenance sources (i.e. same time and network location of the GraphSources g1 and g2 from Algorithm 1), then equal ipv4 address is sufficient for seeding. Next, we iteratively run the Similarity Flooding process until a stable state is reached, typically requiring only a few iterations (see Line 9 and Figure 1, right).

Final steps of the algorithm are on lines 10—13. For each entity type we calculate a relative threshold and output the results (to RDF and GraphML). Note this step encodes another domain-specific assumption: each entity type has an independent similarity threshold. For example, for a given dataset of GraphSources the CNTFO *Router* type has a threshold for similarity that is independent of other types such as *Interface*. We tested three different threshold methods: average, difference and relative.

The average method simply averages all similarity measures for each pair of possible duplicates of one rdf:type. For example in Fig. 1, the average threshold for rdf:type interface pairs is 0.6—the average of *I1-A,I1-B* similarity = 0.2 and *I2-A,I1-B* similarity = 1.0. This method performs the worst because most pairs are not duplicates; i.e. high influence of many near zero-similarity neighbours.

The difference method performs second best. Given an rdf:type, we order all similarities of each possible duplicate pair in ascending order, find the largest difference in similarities between two sequential pairs, then assign the lower-bound of the gap to the threshold.

Marginally performing the best is the relative method. This technique uses the similarity score that is flooded to each rdf:type pair nodes as the threshold. For example the *rdf:type Interface,rdf:type Interface* of Fig. 1 has similarity 0.65 and is assigned as the threshold for rdf:type Interface pairs. This is possible because the rdf:type nodes are also run through the Similarity Flooding iterative process from Algorithm 1.

After duplicate pairs are found, the similarity measures are normalised according to the threshold number chosen. Normalisation is between 0 and 1.0. This ensures that the similarity measures are consistent across all graph comparisons.

4 Results

When interpreting entity resolution results, it is important to note that different domains will have different priorities and consequences for False Positives (FP) and False Negatives (FN). Given that CNTFO is knowledge representation approach, incorrectly classified duplicates (FP) may result contamination of information fusion and subsequent automated reasoning. However, undetected duplicates (FN) may result in unnecessary inferences, as well as reduced information fusion and automated reasoning potential. In our application—Network Path Estimation—we prefer fewer FP and hence a higher precision. This is because factors that influence network paths are already complex to comprehend by analysts, so we strive not to introduce additional confusion due to false network topology induced by false matches. Another important factor when interpreting entity resolution results is the level of certainty. Since we are concerned with exploiting information from potentially untrusted and uncertain datasets, techniques that can provide a indication of similarity and confidence in resolutions are important, which may help manage uncertainty.

We tested our Prolog-based implementation of Similarity Flooding on the ISPNet dataset [14], the same one used in our deterministic approach from [4]. Before generating results, we updated the deterministic approach to consider more entity attributes—to provide a fair comparison. This results in higher accuracy and recall than reported in [4]. Note that a side effect of updated attributes is a lower precision, due to four false positives; i.e. more incorrectly classified duplicates. Also note that the deterministic approach and similarity flooding both used the same attributes, with the deterministic approach adopting rules and the similarity flooding approach adopting comparison attributes for initial seeding.

Table 1 shows the deterministic approach from [4] compared to the three different threshold methods of similarity flooding—relative, difference and average. The relative threshold method not only outperformed the other two thresholding methods, but also outperformed the deterministic approach. The difference threshold shared the same two false positives as the relative threshold and the deterministic approach, with the deterministic approach having two more false positives. This is reflected in the similar precision of the relative and difference threshold methods, and the slightly lower precision for the deterministic app-roach. However, the difference threshold did not find as many duplicate matches, resulting in fewer true positives, as shown by the lower recall. On the other hand, the average threshold method achieved the highest number of true duplicate matches and also the highest number of false duplicate matches, as shown by the recall and precision score.

The false positives in the deterministic approach and the similarity flooding with relative and difference thresholds were due to a dataset error—two routers mistakenly having the same Router ID as two other routers. This caused the deterministic approach to determine, without question, that two pairs were incor-rectly duplicates. On the other hand, while the similarity flooding method also mistakenly indicated that one pair of routers were duplicates, the low similarity rating indicates uncertainty about the decision. Furthermore, there was only one connection between these two routers, which further lowers the certainty. This insight into uncertainty is critical when dealing with unreliable data.

Table 1. Results from similarity flooding and deterministic approaches.

	Deterministic	Similarity flooding (Relative)	Similarity flooding (Difference)	Similarity flooding (Average)
Accuracy	0.9965	0.9966	0.9897	0.9109
Precision	0.9989	0.9995	0.9994	0.5690
Recall	0.9711	0.9711	0.9124	0.9797
F1	0.9848	0.9851	0.9539	0.7199

Future work includes performance validation of our approach against larger datasets and other Entity Resolution techniques. In terms of time performance, our current algorithm requires optimisation to reduce time complexity. In terms of comparing to other techniques, ours exploits domain-specific knowledge and context from CNTFO and measurements to filter comparisons, seed the pair-wise connectivity graph and to provide targeted thresholds for each entity type. Future work includes comparing with techniques that require only limited prior knowledge. We also want to investigate neighbour influence in the similarity flooding algorithm. Should one neighbour among five have less influence than a sole neighbour? Should different relationships or literals have greater weight-ings than others due to their domain-specific importance? Should initial seeding values have less influence? Another possible research question could investigate how to merge data from sources with fluctuating reliability (see [15]).

5   Path Estimation and Ranking

Using similarity measures we can rank network paths that contain duplicates
with highest certainty resolutions at the top. When making decisions about
preferred best network paths, this ranking provides insight into the inherent
uncertainty in the underlying data.

Figure 2 provides an example of path estimation across three GraphSources—
CORE, TRACEROUTE and ARPING. These GraphSources have different
provenance and different levels of uncertainty (due to varying reliability) in relation to the path estimation problem. The CORE GraphSource (orange, top)
provides relatively complete and detailed information, though typically it is less
timely and does not capture dynamic changes in the network. The TRACER-
OUTE GraphSource (green, middle) is a small subset of dynamic network information that is typically timely and brief, but lacks detail relative to the CORE
GraphSource. The ARPING GraphSource (blue, bottom) is likewise a small subset of higher detail information, but lacks breadth of information relative to the
CORE GraphSource. Because all three GraphSources provide information about
overlapping sets of network elements, they are excellent candidates for information fusion. However these overlaps lead to duplicated entities, and hence Entity
Resolution is needed for information fusion to occur. Each owl:sameAs link in
Fig. 2 (shown in red) corresponds to normalised similarity measure, as discussed
in Sect. 3. The similarity measure on each link be used to construct a graph
and assign a distance value on each vertex. By fixing the source and destination
node, these distance values can be used by Dijkstra's shortest path algorithm
to find and rank paths by total distance[1]. The three ranked paths are indicated
by green, blue and orange glowing lines in Fig. 2, starting in C1-MEL (top left)
and ending in C1-ADL (top right). In that work, we showed how duplicates lead
to multiple estimated paths, with entity resolution resulting in a set of recommended path but lacking quantification of reliability in the resulting paths. Here

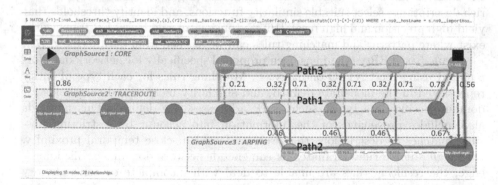

Fig. 2. Entity links across 3 GraphSources—CORE, TRACEROUTE and ARPING –
with owl:sameAs link similarity measures in red, which can be rank paths as shown
(Color figure online)

<hr>

[1] We computed ranked paths using Neo4j's All Shortest Paths algorithm [1].

in Fig. 2 we show that because the path crosses multiple GraphSources — each with different provenance and uncertainty—we can recommend *three* paths in ranked order. This ranking is important for decision making. Lower ranked paths are relatively less certain, with higher ranked paths more likely to correspond to paths traversed by real-time network traffic. The question *"Which is the best path?"* is then answered with higher confidence and insight into uncertainty. Ranked paths also become additional semantics with attached provenance, further enriching the results.

6 Related Work

Many approaches to entity resolution (ER), data linking, and data de-duplication exist in the literature, with application in numerous scientific fields. ER techniques can be broadly distinguished based on the nature of similarity measure and whether the method makes use of (usually relational model) data schemas that may govern the structure of the data; the nature of matching entities (pairwise or collectively); and the method of processing (offline or online) [3]. In general, structural, linguistic, and statistical properties of set of entity descriptions are considered simultaneously to compute similarity scores in modern matching systems. In the following we highlight selected relevant works and refer the readers to recent survey articles for a more general overview [3,5,6,9].

Rule-based approaches to ER are commonly found in the database literature, as rules offer the ability to explicitly encode domain knowledge into the resolution process. Rules also assist explanation and debugging. At the core of rule-based systems are rules of the form *"if two entity descriptions a and b are 'similar', then the a and b refer to the same entity"*. The rules encode which attributes of the descriptions are relevant, and similarity functions quantify how similar the values of an attribute present in both descriptions are. Unfortunately, rule-based ER systems can be brittle as it can be difficult to generate matching rules for rich descriptions, select suitable similarity functions, and identify thresholds that yield high precision and high recall [16]. Extensions of rule-based ER frameworks exist where patterns and matching rules can be learned from data (see e.g. [7]).

A rule-based ER method has been proposed specifically for network data formulated using the CNTFO ontology [4]. Matching rules for each type of entity represented in the model are defined, and similarity metrics and thresholds for domain-specific data types, such as IP addresses, are defined. This methods also includes provenance of the information into account by restricting possible matches to descriptions that were collected in close temporal proximity. Although tailored rules for matching can encode network domain specific knowledge, the resulting rules are unable to deal with incomplete or contradictory information that may be found in network traces.

In contrast to schema-informed methods, techniques for schema-less ER usually rely on a combination of attribute value similarity and structural similarity of graphs connecting entity description (and their parts). Whereas attributes identify candidate correspondences between vertices in the graphs, the connections in the graph structures facilitate the propagation and aggregation of similarities for entire sub-structures.

Similarity flooding [8] is a well-known approach to matching based on ideas from percolation theory, where a set of corresponding entity descriptions is identified initially, and neighbouring entity descriptions influence each others' similarity scores in an iterative approach until a stable state is reached. This principle has been adopted in schema matching and ontology alignment [10,11]. In contrast to rule based ER, similarity flooding does not require domain-specific matching rules, it can arrive at meaningful results if contradictory information is present, and the propagation-based ideas naturally ensure consistency of similarity scores along paths connecting entities.

In this work we extended similarity flooding with domain specific rules that govern which representations can and cannot form correspondences (akin to type-based blocking approaches that are at the core of most ER systems [3]), and investigated different approaches to define thresholds for extracting matches based on collective similarity scores.

Graph-based representations of entities in the context of Linked Open Data and RDF pose additional challenges to ER, as the representation of the same entities in different graphs may differ syntactically and structurally [2]. For graph representations that follow similar naming conventions, matching can be performed based on the components of the identifiers of the descriptions [3]. For heterogeneous graphs, entity representations can be distilled in to collections of tokens, akin to words in a sentence, and techniques from information retrieval and natural language processing can be applied to score similarity. For example, [2] transform entity descriptions into a vectors that can subsequently be clustered to identify potential matches. New links (e.g. *rdfs:sameAs*) are subsequently created between matching entities and added to the graphs. Nentwig et al. [9] presents a survey of tools for link discovery in RDF data sets.

Machine learning techniques have been adopted for data integration and ER in recent years. Here, learning methods have been applied to detect matches at the schema level, infer matching rules and thresholds, quantify the quality of different sources and their impact on the confidence of matches. These issues deserve attention in particular in contexts such as information extraction from the Web, where a wide array of different sources must be reconciled. We refer the reader to [5] for an overview of relevant works in this area.

Dense vector representations of graph substructures via graph embeddings are becoming increasingly popular for data integration and ER, as embeddings enable to learn and assess the semantic similarity of nodes, relationships, and entire substructures based on latent properties that are learned from examples. Iterating embedding and alignment operations can derive alignments between multiple heterogeneous graphs structures [17].

In our context, the heterogeneity is controlled by the overarching CNTFO ontology, which defines the main concepts and attributes used throughout the representations of entities in different data sets. However, machine learning techniques will be an important next step for reliable ER where incomplete and conflicting information over time are prevalent, for example when relying on untrusted third-party data. Developing suitable ER and network data fusion methods is subject to future work.

7   Conclusion

This paper presents an approach to Similarity Flooding to improve network path estimation. Our approach outperforms previous deterministic rule-based approach on uncertain data. Domain-specific knowledge can be incorporated through ontologies, helping filter unnecessary comparisons, seed the pairwise connectivity graph and to provide targeted thresholds for each entity type. However, the presented approach is independent from the specific ontology and can extend to other domains where different feature combinations, weights, and thresholds may apply. Using similarity measures between duplicates we can rank network paths that contain duplicates with highest certainty resolutions at the top. We showed that path ranking is important for decision making—ranks indicating level of certainty. Higher ranked paths are more likely to correspond those traversed by real-time network traffic, which is important for cyber situational awareness. The question *"Which is the best path?"* is then answered with higher confidence and deeper insight into uncertainty in the underlying data that is increasingly context-sensitive, unreliable and heterogeneous. Furthermore, ranked paths can be piped into subsequent automated reasoning agents in an artificial intelligence application. Provenance of each ranked path informs subsequent agents that may be ingesting ranked paths from multiple sources. This ranking and provenance-chain also contributes to trusted explainable-AI.

References

1. Neo4j graph database management system. https://neo4j.com/. Accessed 15 July 2010
2. Achichi, M., Bellahsene, Z., Ellefi, M.B., Todorov, K.: Linking and disambiguating entities across heterogeneous RDF graphs. J. Web Semant. **55**, 108–121 (2019). https://doi.org/10.1016/j.websem.2018.12.003
3. Christophides, V., Efthymiou, V., Palpanas, T., Papadakis, G., Stefanidis, K.: End-to-end entity resolution for big data: A survey. CoRR abs/1905.06397 (2019). http://arxiv.org/abs/1905.06397
4. Philp, D., Chan, N., Sikos, L.F.: Decision support for network path estimation via automated reasoning. In: Czarnowski, I., Howlett, R.J., Jain, L.C. (eds.) Intelligent Decision Technologies 2019. SIST, vol. 142, pp. 335–344. Springer, Singapore (2020). https://doi.org/10.1007/978-981-13-8311-3_29. Chap. 29
5. Dong, L., Rekatsinas, T.: Data integration and machine learning: a natural synergy. PVLDB **11**(12), 2094–2097 (2018). https://doi.org/10.14778/3229863.3229876. www.vldb.org/pvldb/vol11/p2094-dong.pdf
6. Dorneles, C.F., Gonçalves, R., dos Santos Mello, R.: Approximate data instance matching: a survey. Knowl. Inf. Syst. **27**(1), 1–21 (2011). https://doi.org/10.1007/s10115-010-0285-0

7. Kwashie, S., Liu, J., Li, J., Liu, L., Stumptner, M., Yang, L.: Certus: an effective entity resolution approach with graph differential dependencies (gdds). PVLDB **12**(6), 653–666 (2019). www.vldb.org/pvldb/vol12/p653-kwashie.pdf
8. Melnik, S., Garcia-Molina, H., Rahm, E.: Similarity flooding: a versatile graph matching algorithm and its application to schema matching. In: Proceedings 18th International Conference on Data Engineering, pp. 117–128. IEEE (2002)
9. Nentwig, M., Hartung, M., Ngomo, A.N., Rahm, E.: A survey of current link discovery frameworks. Semant. Web **8**(3), 419–436 (2017). https://doi.org/10.3233/SW-150210
10. Rahm, E., Bernstein, P.A.: A survey of approaches to automatic schema matching. VLDB J. **10**(4), 334–350 (2001). https://doi.org/10.1007/s007780100057
11. Shvaiko, P., Euzenat, J.: Ontology matching: state of the art and future challenges. IEEE Trans. Knowl. Data Eng. **25**(1), 158–176 (2013). https://doi.org/10.1109/TKDE.2011.253
12. Sikos, L.F., Stumptner, M., Mayer, W., Howard, C., Voigt, S., Philp, D.: Automated reasoning over provenance-aware communication network knowledge in support of cyber-situational awareness. In: Liu, W., Giunchiglia, F., Yang, B. (eds.) Knowledge Science, Engineering and Management, pp. 132–143. Springer International Publishing, Cham (2018). https://doi.org/10.1007/978-3-319-99247-1_12
13. Sikos, L.F., Stumptner, M., Mayer, W., Howard, C., Voigt, S., Philp, D.: Representing network knowledge using provenance-aware formalisms for cyber-situational awareness. Proc. Comput. Sci. **126**, 29–38 (2018)
14. Sikos, L.F., Stumptner, M., Mayer, W., Philp, M.D., Voigt, M.S., Howard, C.: Provenance-aware lod datasets for detecting network inconsistencies. In: CEUR Workshop Proceedings of Contextualized Knowledge Graphs and Semantic Statistics, vol. 2317 (2018)
15. To, A., Meymandpour, R., Davis, J.G., Jourjon, G., Chan, J.: A linked data quality assessment framework for network data. In: Proceedings of the 2nd Joint International Workshop on Graph Data Management Experiences & Systems (GRADES) and Network Data Analytics (NDA), GRADES-NDA 2019, pp. 4:1–4:8. ACM, New York (2019). https://doi.org/10.1145/3327964.3328493
16. Wang, J., Li, G., Yu, J.X., Feng, J.: Entity matching: how similar is similar. PVLDB **4**(10), 622–633 (2011). https://doi.org/10.14778/2021017.2021020. www.vldb.org/pvldb/vol4/p622-wang.pdf
17. Zhu, H., Xie, R., Liu, Z., Sun, M.: Iterative entity alignment via joint knowledge embeddings. In: Sierra, C. (ed.) Proceedings of the Twenty-Sixth International Joint Conference on Artificial Intelligence, IJCAI 2017, pp. 4258–4264, Melbourne, Australia, 19–25 August 2017 (2017). https://doi.org/10.24963/ijcai.2017/595

Interactive Deep Metric Learning for Healthcare Cohort Discovery

Yang Wang[1]([✉]), Guodong Long[1]([✉]), Xueping Peng[1]([✉]), Allison Clarke[2], Robin Stevenson[2], and Leah Gerrard[2]

[1] Centre for Artificial Intelligence, University of Technology Sydney, Sydney, NSW, Australia
Yang.Wang-17@student.uts.edu.au, {Guodong.Long,Xueping.Peng}@uts.edu.au
[2] Department of Health, Australian Government, Canberra, ACT, Australia
{Allison.Clarke,Robin.Stevenson,Leah.Gerrard}@health.gov.au

Abstract. Given the continuous growth of large-scale complex electronic healthcare data, a data-driven healthcare cohort discovery facilitated by machine learning tools with domain expert knowledge is required to gain further insights of the healthcare system. Specifically, clustering plays a crucial role in healthcare cohort discovery, and metric learning is able to incorporate expert feedback to generate more fit-for-purpose clustering outputs. However, most of the existing metric learning methods assume all labelled instances already pre-exists, which is not always true in real-world applications. In addition, big data in healthcare also brings new challenges to metric learning on handling complex structured data. In this paper, we propose a novel systematic method, namely Interactive Deep Metric Learning (IDML), which uses an interactive process to iteratively incorporate feedback from domain experts to identify cohorts that are more relevant to a particular pre-defined purpose. Moreover, the proposed method leverages powerful deep learning-based embedding techniques to incrementally gain effective representations for the complex structures inherit in patient journey data. We experimentally evaluate the effectiveness of the proposed IDML using two public healthcare datasets. The proposed method has also been implemented into an interactive cohort discovery tool for a real-world application in healthcare.

Keywords: Clustering · Deep metric learning · Interactive cohort discovery · Patient journey similarity

1 Introduction

With the recent growing adoption of Electronic Health Record (EHR), more large-scale electronic datasets are collected from healthcare systems. To enable data or policy analysts to gain a better understanding of big data in healthcare, a data-driven analysis tool that could assist analysts to discover the cohorts according to patients' EHRs is needed. With the discovered cohorts, groups of

T. D. Le et al. (Eds.): AusDM 2019, CCIS 1127, pp. 208–221, 2019.
https://doi.org/10.1007/978-981-15-1699-3_17

patients with certain "similarity" in health records will be presented to the analysts for further analysis. However, how to define the "similarity" is a practical challenge. In particular, there are three major issues that need to be considered before the "similarity" can be defined, including (1) different analysis purposes may require different similarity definition, (2) different factors in the EHR may have different impacts to the similarity measurement, and (3) the mixture of various data types may increase the difficulty of data pre-processing.

Given an EHR dataset, various analysis tasks can be derived and conducted, e.g. healthcare resource utilisation, treatment comparison, or re-admission analysis. Clustering plays a significant role in all these tasks. Therefore, a fit-for-purpose clustering approach is needed to incorporate generalised clustering algorithms and experts' domain knowledge related to the purpose. Metric learning is one of the machine learning approaches to transform the expert's feedback into a fit-for-purpose similarity measurement. In particular, it will adjust similarity by finding a set of optimal weights for the features that maximally fits expert's feedback.

Although many traditional metric learning methods have been successfully applied to healthcare datasets, most of them are based on an implicit assumption that all the labelled instances are provided in advance and at a same time. However, in many real-world applications, constraints or labels are often available only incrementally, and it is difficult to determine what and how much labelling information is needed for a fit-for-purpose model. In addition, many domain experts prefer a simple way to perform labelling: such as labelling based on the visualisations of each iteration of clustering result. Unfortunately, traditional metric learning methods are insufficient in these circumstances.

In addition, many metric learning applications in healthcare are based on handcrafted feature representations, which have multiple limitations and cannot generate an effective and comparable representation for each patient. Recently, deep metric learning approaches have been proposed to deal with these limitations. For instance, Suo et al. [26] proposed a deep metric learning framework on EHR data that derived effective representations for patients without loss of their historical and temporal information and then calculated the similarity measurement based on these learned representations. Similarly, Miotto et al. [22] applied a deep metric learning approach to derive a general-purpose patient representation from EHR data that facilitated clinical predictive modeling. However, these new frameworks cannot handle a continuous stream of expert feedback.

To tackle these various challenges, this paper propose a novel model, called Interactive Deep Metric Learning (IDML). This utilises Recurrent Neural Network (RNN) based embedding techniques to learn the relationships among individual medical concepts and healthcare encounters, metric learning to incorporate expert knowledge into the construction of a customised distance metric, and an interactive approach to handle the continuous stream of expert feedback and to enable the expert to review and revise the model iteratively without requiring programming knowledge. Our contributions are as follows:

- This paper proposes a new practical framework for clustering. This is a systematic method that solves existing practical challenges by integrating multiple machine learning procedures.
- This paper is a new application for healthcare cohort discovery. In particular, a tool was deployed in the Australian Government Department of Health for fine-grained policy analysis.
- The proposed method is an effective framework that is demonstrated by two publicly available datasets.

The remainder of this paper is organized as follows: Sect. 2 briefly reviews some related work, with special focus on the work that has been applied to EHR data. Section 3 describes the technical details of the proposed IDML model. Section 4 presents experimental results on two public datasets. Section 5 concludes by summarizing our main contributions. Finally Appendixes provide additional notes and computational formulas that are related to this research.

2   Related Work

This section briefly reviews some related work, with special focus on the work that has been applied to EHR data.

2.1   Healthcare Cohort Discovery

The improvement and increase in use of technology has led to the wide adoption of electronic datasets collected by healthcare systems [25]. The analysis and interpretation of these healthcare datasets is essential to develop high quality healthcare, improved healthcare management, reduced healthcare costs, population health management and effective clinical research [19]. Healthcare cohort discovery is an important part of healthcare data analysis. In particular, given the fact that policy effectiveness is partially determined by ability to target the right people [5], fit-for-purpose patient cohort discovery has the potential to play a significant role in effective healthcare policy making. Take cohort discovery in patient journey for instance, to obtain optimal outcomes for patients and assist in allocating healthcare resources, it is essential that policies effectively targeted patient groups that are most likely to benefit. An effective patient cohort discovery by a fit-for-purpose patient journey clustering allows identification of patients who are similar or different from one another, or from typical journeys, or from best practice clinical pathways. This can provide insight to drive exploration of patients who are dissimilar from their peers and may achieve better health outcomes from additional health services or support. However, it is always challenging to devise a fit-for-purpose metric for cohort discovery or clustering which is relevant to different clinical or policy contexts [29].

2.2   Interactive Metric Learning

Distance metric learning has been proposed to incorporate expert knowledge into the construction of a customised distance metric, which is required for a fit-for-purpose cohort discovery. More specifically, distance metric learning algorithms can be classified as unsupervised [13,17], semi-supervised [27,33] or supervised [12,30], based on the availability of label or constraint information in the training dataset. In particular, semi-supervised and supervised distance metric learning constructs a distance metric that maps similar samples closer to each other, while those that are dissimilar are mapped further apart from each other, which is particularly relevant for patient similarity learning. This is because this kind of metric learning algorithm provides a natural way to incorporate feedback from domain experts [29].

To make the distance metric learning more applicable to real-world applications, it is necessary to be able to efficiently and effectively incorporate new feedback into the existing model. In other words, the learned distance metric needs to be incrementally updated without expensive rebuilding [29]. Wang and Sun [28] agreed on this and introduced an interactive patient similarity framework to handle a continuous stream of expert feedback and to make use of not only supervised information (i.e., limited physician feedback) but also unsupervised information (e.g., patient features similarity).

2.3   Deep Learning-Based Embedding Techniques

EHR data is challenging to represent and model due to its high dimensionality, noise, heterogeneity, sparseness, incompleteness, random errors, and systematic biases [15,31,32], and the same medical concept can also be expressed using different codes and terminologies [22]. This is another major challenge to applying distance metric learning to healthcare applications. To deal with this challenge, researchers have recently begun attempting to apply deep learning methods to EHR data to utilise its ability to learn complex patterns from data [6]. Given that the previous studies such as representation learning [7,9,22], had not fully made use of the sequential nature of medical record data, researchers have started to use RNNs to develop a representation of patient medical histories that incorporat both the temporal characteristics of the data and the relationships among individual clinical concepts [6,18]. Such studies showed RNNs can better represent the longitudinal EHR sequence as their proposed frameworks outperformed many classic "shallow" distance metric learning methods. However, those new frameworks cannot handle a continuous stream of expert feedback.

3   Method

In this section, we give the details of our proposed interactive deep metric learning model for healthcare cohort discovery. We first give several definitions of medical concepts, then show how to learn effective representations for patient journeys in the longitudinal EHR data, and lastly introduce a method that can learn relative similarity between patient pairs.

3.1   Definitions

Each EHR is a sequential record of a patient's healthcare visits, where each visit is logged as a set of medical codes indicating the disease or treatment the patient suffered or received [23, 24, 26]. For the ease of understanding the medical concepts, we first give the definitions of them as follows.

Definition 1 (Medical Code). *A medical code is defined as a term or entry to describe diagnosis, procedure, medication, and laboratory tests for an inpatient during a treatment process. We denote the set of medical codes as $C = \{c_1, c_2, \ldots, c_{|C|}\}$, where $|C|$ is the total number of unique medical codes in the EHRs.*

Definition 2 (Visit). *A visit for an inpatient refers to a treatment process from admission to discharge with an admission time stamp. We denote a visit as $V_{i,t} = [c_1^{i,t}, c_2^{i,t}, \ldots, c_K^{i,t}]$, where i is the i-th patient, t is admission time of the visit of the patient (here we only consider the order of visits, so t can be thought of as the t-th visit), K is the size of medical codes in a visit, and $c_k^{i,t}$ is the k-th medical code.*

Definition 3 (Patient Journey). *A patient journey consists of a sequence of visits over time, which is denoted as $J_i = [V_{i,1}, V_{i,2}, \ldots, V_{i,M}]$, where M is the total visit times for the patient i.*

In the rest of this paper, a patient's medical data is stored to a sequence of M visits in patient journey J_i chronologically. Hence, to reduce clutter, we omit the superscript and subscript (i) indicating i-th patient, when we are discussing a single patient journey.

3.2   Representation Learning

Representation for Medical Code. Medical code embedding is a fundamental processing unit in deep neural network based EHR. It transfers each discrete medical code into distributed real-valued vector representation. Formally, given a sequence or set of medical codes $c = [c_1, c_2, \ldots, c_n] \in \mathbb{R}^{|C| \times n}$, where c_i is a one-hot vector, and n is the sequence length. Typically, in a NLP literature, a word embedding method(e.g. word2vec [20, 21]) is applied to the sequence, which outputs a sequence of low dimensional vectors $e = [e_1, e_2, \ldots, e_n] \in \mathbb{R}^{d \times n}$, where d is the embedding dimension of e_i. This process can be formally written as $e = W^{(e)} c$, where $W^{(e)} \in \mathbb{R}^{d \times |C|}$ is the embedding weight matrix that can be fine-tuned during the training phase.

Representation for Visit. As defined in Definition 2, a visit consists of a set of medical codes. Therefore, a visit can be represented by the set of medical codes occurred in the visit, when each medical code is embedded into a real-valued dense vector. One straightforward way is to sum each embedded medical code in the visit to learn visit representation v_t using the formula:

$$v_t = \sum_{c_i \in V_t} W^{(e)} c_i, \tag{1}$$

where V_t is the set of medical codes in t-th visit, c_i is the i-th code in V_t which is a one-hot vector, and $W^{(e)} \in \mathbb{R}^{d \times |C|}$ is the embedding weight matrix.

Representation for Patient Journey. The visits in a patient journey are sequential but intermittent. To capture the temporal semantic relationships among medical codes, we exploit RNN to learn representations for patient journeys. The learned patient journey embedding contains not only visit information of the patient, but also the sequential relationship among visits.

In this paper, we employ the basic structure of Long Short-Term Memory (LSTM) unit [14] to the patient journey embedding. The detailed LSTM formulation can be found in Appendix A.

3.3 Deep Metric Learning

Learning a fit-for-purpose similarity between each pair of patient journeys is the key step for healthcare cohort discovery.

Metric Learning. We follow the idea of traditional metric learning to learn a fit-for-purpose distance metric for patients. Specifically, in metric learning, a key step is to learn a positive semi-definite matrix A to transform the raw data into a new space. The new metric in the space can better reflect the relationships of uses' given feedback. The distance between two points x_i and x_j can be obtained using following equation,

$$d^2(x_i, x_j) = \|x_i - x_j\|_A^2 = (x_i - x_j)^T A (x_i - x_j), \tag{2}$$

RNN-based Metric Learning. In RNN-based metric learning, the transformation matrix A can be learned by a complex nonlinear neural network. In the task of cohort discovery, the patient similarity means to measure the distance among different patient journeys that consist of sequences of visits with time steps. RNN is an appropriate neural network that can be utilised to represent such sequences and learn the nonlinear transformation as follows,

$$d^2(p_i, p_j) = \|R(p_i) - R(p_j)\|^2 = \|h_i - h_j\|^2, \tag{3}$$

where R is the RNN operation in our task setting, and h_i and h_j are the representations for patient journeys p_i and p_j learned via Eq. 6.

In order to ensure that patients from the same class are closer to each other, while those from different classes are further apart from each other, we have the objective function as follows:

$$\mathcal{L} = -(y \log(d^2(p_i, p_j)) + (1 - y) \log(d^2(p_i, p_j))) \tag{4}$$

$$y = \begin{cases} 0 \text{ for } (p_i, p_j) \in S \\ 1 \text{ for } (p_i, p_j) \in D \end{cases} \tag{5}$$

where $(p_i, p_j) \in S$ means p_i and p_i are similar and $(p_i, p_j) \in D$ means p_i and p_i are dissimilar.

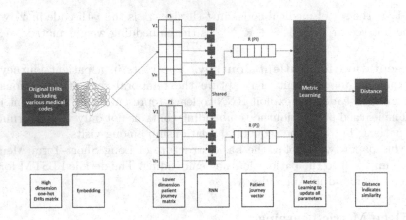

Fig. 1. The framework of RNN based deep metric learning neural network

The proposed metric learning is added on top of the RNN, which takes the learned vector representation as the input to calculate distance between patients. The objective function above is minimized through back propagation, and all the parameters are updated simultaneously. The learned distance metric indicates the similarity between patient pairs, with smaller distance values for higher similarity. The framework of RNN based deep metric learning is shown in Fig. 1, which is also an end-to-end learning framework.

3.4   Interactive Cohort Discovery

To improve the applicability and efficiency of metric learning, we integrate an interactive approach and visualisation functions into the proposed IDML to allows domain expertise to be incorporated effectively, interactively, and also iteratively.

Interactive clustering models [3,4] were inspired by an analogous model for learning under feedback [1]. In the model, the clustering algorithm proposes a hypothesis to the user and obtains feedback regarding the suitability of the current hypothesis [2]. Similarly, in this paper, we introduce an interactive framework to handle the continuous stream of expert feedback and to enable expert to review outputs and revise the patient cohort discovery model iteratively. Specifically, in the proposed IDML we provide visualisations of each iteration of clustering result, which enables experts to identify the patients who most need to be investigated or labelled. Based on the proposed IDML, we have also developed an interactive cohort discovery tool which has been deployed in the Australian Government Department of Health for future exploration.

4 Experiments

We evaluate our proposed model on two public datasets by clustering tasks.

4.1 Data Description

The experiments are based on two public datasets - **MIMIC III** [16] and **CMS** [10]. To perform our research, we extract three patient cohorts from the dataset: diabetes, chronic obstructive pulmonary disease (COPD) and heart failure (HF). Following the disease selection criteria in [8,26], we identify the diseased patients who have (1) qualifying ICD-9 codes for a specific disease in the encounter records, and (2) at least two and five clinical encounters with qualifying ICD-9 codes occur in **MIMIC III** and **CMS**, respectively. The date when the first target diagnosis code appears is denoted as the decision date. We split the patient sequences at the decision date into two parts, and use only the part before the decision date which contains early symptoms and complications for similarity learning. To enable distinct cohorts, we remove overlapped patients so that each patient only suffers from one disease. The statistical information for both datasets is listed in Table 1.

Table 1. Statistics of datasets

Cohorts	MIMIC III			CMS		
	Diabetes	COPD	HF	Diabetes	COPD	HF
# of patients	205	170	448	1,940	1,934	2,124
# of visits	632	523	1,377	12,134	12,067	13,220
Avg. # of visits per patient	3.08	3.08	3.07	6.25	6.24	6.22
# of unique diagnose codes	1,255	1,167	1,698	3,431	3,480	3,542
# of unique procedure codes	385	350	509	2,143	2,131	2,277

4.2 Experimental Setup

The proposed model and baseline machine learning models were implemented in the Python 3.7 environment. To build the deep learning related models, we applied TensorFlow (2.0.0-rc1) backend. For the traditional models, we adopted the models from Sklearn 0.20.2.

All the experiments were conducted on a workstation with 10 cores of Intel Core i9 CPU and 64 GB RAM. Due to the high computation cost of deep learning algorithm, we configured one GPU (Nvidia TITAN Xp, 12 GB RAM) to accelerate the training speed of the models.

In this experiment for our proposed model IDML, we set 100 as the embedding dimension of medical code and 5 rounds of interactive iteration with 100 labelled pairs of patients (half labelled as similar pairs and the other half labelled as different pairs) being provided in each round. Specially, to emulate human behaviors, all the labelled pairs are only comprised of the patients who are near the edge of one cluster and also close to a neighbouring cluster in each round.

To make a fair comparison, we repeat all the methods 50 times using random initialization and record the average result. We also make all methods involving metric learning incorporate the same amount (i.e. 500) of labelled pairs in total.

Baseline Approaches. We compare the proposed *IDML* with the following popular baseline methods:

Euclidean: In this basic metric approach, we first represent a patient as a one-hot vector, whose dimension is $|C|$, and the corresponding entry is 1 if the medical code is occurred in a patient journey. Then, we apply Principal Component Analysis (PCA) to reduce dimension of the one-hot vector to a low-dimensional vector, whose dimension is 100. Finally, we apply Euclidean distance to similarity calculation and perform clustering in the low-dimensional space.

ITML: Information-Theoretic Metric Learning, a traditional metric learning method which learns a Mahalanobis distance by minimizing the differential relative entropy between two multivariate Gaussians under constraints on the distance function [11], is applied to the same low-dimensional vectors prepared in Euclidean method. In this method, we provide 500 labelled pairs of patients (half labelled as similar pairs and the other half labelled as different pairs) in a **batch** to train a new metric.

DML: Deep Metric Learning method mentioned in Sect. 3.3 has been applied. Same with the labelling strategy for ITML, we also provide 500 labelled pairs of patients with half labelled as similar and the other half labelled as different in a **batch** to train a new metric.

Clustering Evaluation Measures. Clustering results are evaluated by three widely used criteria: *Normalized Mutual Information (NMI)*, *Adjusted Rand Index (ARI)*, and *Purity*. A detailed description of these three measures is provided in Appendix B.

4.3 Results

This section presents the clustering results of all the above methods. We used K-Means as the clustering algorithm and evaluated the performance by the three measures mentioned above (i.e. NMI, ARI and Purity).

Overall Performance. Table 2 demonstrates the clustering performance evaluated by the above three measures. The best results appear in **bold**. As we can see from the table, our proposed IDML outperforms two baseline methods (i.e. Euclidean and ITML) and also beats its non-interactive counterpart (i.e.

DML) on both MIMIC III and CMS datasets. Based on this, we draw the conclusions: (1) metric learning helps to improve clustering performance, (2) deep metric learning models obtain better performance than traditional ones, and (3) interactive model is better than any non-interactive ones. Moreover, we find all models perform better when based on MIMIC III dataset than on CMS. This may because MIMIC III is a real-world dataset and has a smaller size than CMS.

Table 2. Clustering performance comparison

Methods	MIMIC III			CMS		
	NMI	ARI	Purity	NMI	ARI	Purity
Euclidean	0.0031	0.0087	0.5443	0.0004	0.0001	0.3541
ITML	0.0043	0.0183	0.5443	0.0007	0.0004	0.3550
DML	0.0320	0.0122	0.5443	0.0019	0.0014	0.3613
IDML	**0.0343**	**0.0378**	**0.5480**	**0.0025**	**0.0021**	**0.3701**

Influence of Interactive Iteration. Figure 2(a) and (b) indicate the clustering performance of proposed IDML at different interactive rounds, where the X-axis corresponds to the feedback rounds ranging from 1 to 5, and the Y-axes represent the values of the three measures mentioned above. Specially, a secondary Y-axis on the right side has been added to handle the different scales among the three measures (the primary Y-axis shows the values of NMI and ARI, and the secondary Y-axis shows the Purity value). These figures show that the clustering performance of IDML on both MIMIC III and CMS datasets generally increases with the sequential update process.

Particularly, both NMI and ARI values in MIMIC III and CMS datasets gradually grow with the increase of the interactive rounds, which demonstrates that our proposed model is able to integrate domain experts' knowledge round by round to develop more fit-for-purpose results. For the Purity value, it is stable by varying round from 1 to 4 in MIMIC III, but jumps to highest point at round 5; While in CMS, the Purity value gradually increases when interactive round changes from 1 to 3, but it moves back to a lower point at round 4 and jumps to the largest value at 5. In general, the evaluation measure Purity also reflects our proposed model working well.

5   Conclusion

In this paper we have proposed a novel systematic method, namely Interactive Deep Metric Learning (IDML), for healthcare cohort discovery. The new framework leverages deep metric learning to build a more comprehensive representation for patient journeys that can then utilise adapted similarity measures

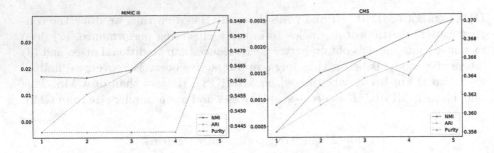

Fig. 2. IDML performance at different feedback rounds

to produce effective and relevant clusters. In addition, taking into account the real-world situation and health experts' preference, we have integrated an interactive approach with visualisation functions into the framework to handle the continuous stream of expert feedback and to enable experts to review and revise the model iteratively. Experiments using two public healthcare datasets have demonstrated the effectiveness of this framework. We have also implemented the framework into an interactive cohort discovery tool for a real-world application in healthcare.

A   LSTM Formulation

$$
\begin{aligned}
i_t &= \sigma(W^i v_t + U^i h_{t-1}) \\
f_t &= \sigma(W^f v_t + U^f h_{t-1}) \\
o_t &= \sigma(W^o v_t + U^o h_{t-1}) \\
\tilde{c}_t &= tanh(W^c v_t + U^c h_{t-1}) \\
c_t &= f_t * c_{t-1} + i_t * \tilde{c}_t \\
h_t &= tanh(c_t) * o_t
\end{aligned}
\tag{6}
$$

where i_t, f_t, o_t are input, forget and output gates respectively. The gates are different neural networks that decide which information is allowed on the cell state. The gates can learn what information is relevant to keep or forget during training. v_t is an input to a network at time t, h_{t-1} is an output at time $t-1$ and c_{t-1} is an internal cell state at $t-1$.

B   Clustering Evaluation Measures

- The *Normalized Mutual Information (NMI)* is defined as:

$$
NMI(\widehat{K}; K) = \frac{2 \times I(\widehat{K}; K)}{\left[H(\widehat{K}) + H(K)\right]}
\tag{7}
$$

where $I(\widehat{K}; K)$ is the mutual information and the entropies $H(\widehat{K})$ and $H(K)$ are used for normalizing the mutual information to be in the range of $[0, 1]$. The higher the NMI is, the better the corresponding clustering is.

- The *Adjusted Rand Index (ARI)* of clustering is defined as:

$$ARI = \frac{RI - E[RI]}{max(RI) - E[RI]} \tag{8}$$

where $RI = \dfrac{a+b}{C_2^N}$, a is the number of patient pairs coming from the same cohort and also grouped into same cluster, b is the number of patient pairs belonging to different cohorts and grouped into different clusters. N is the total number of patients. The range of ARI values is $[-1, 1]$. The higher the ARI is, the better the corresponding clustering is.

- The *Purity* of clustering is defined as:

$$Purity(\widehat{K}, K) = \frac{1}{N} \sum_{i=1}^{|\widehat{K}|} \max_j \left| \widehat{k_i} \cap k_j \right|, \tag{9}$$

where $\widehat{K} = \{\widehat{k_1}, \widehat{k_2}, \ldots, \widehat{k_{|\widehat{K}|}}\}$ is the set of clusters produced by the chosen clustering algorithms. $|\widehat{K}|$ is the total number of clusters. $K = \{k_1, k_2, \ldots, k_{|K|}\}$ is the group of patient cohorts (i.e. ground truth). $\max_j \left| \widehat{k_i} \cap k_j \right|$ is the size of the intersection of cluster $\widehat{k_i}$ and patient cohort k_j which is most frequent inside. The range of *Purity* values is $[0, 1]$. The higher the *Purity* is, the better the corresponding clustering is.

References

1. Angluin, D.: Queries and concept learning. Mach. Learn. **2**(4), 319–342 (1988)
2. Awasthi, P., Balcan, M.F., Voevodski, K.: Local algorithms for interactive clustering. J. Mach. Learn. Res. **18**(1), 75–109 (2017)
3. Balcan, M.-F., Blum, A.: Clustering with interactive feedback. In: Freund, Y., Györfi, L., Turán, G., Zeugmann, T. (eds.) ALT 2008. LNCS (LNAI), vol. 5254, pp. 316–328. Springer, Heidelberg (2008). https://doi.org/10.1007/978-3-540-87987-9_27
4. Balcan, M.F., Liang, Y., Gupta, P.: Robust hierarchical clustering. J. Mach. Learn. Res. **15**(1), 3831–3871 (2014)
5. Brainard, W.C.: Uncertainty and the effectiveness of policy. Am. Econ. Rev. **57**(2), 411–425 (1967)
6. Choi, E., et al.: Doctor AI: predicting clinical events via recurrent neural networks. In: Machine Learning for Healthcare Conference, pp. 301–318 (2016)
7. Choi, E., et al.: Multi-layer representation learning for medical concepts. In: SIGKDD, pp. 1495–1504. ACM (2016)
8. Choi, E., et al.: RETAIN: an interpretable predictive model for healthcare using reverse time attention mechanism. In: NIPS, pp. 3504–3512 (2016)
9. Choi, Y., Chiu, C.Y.I., Sontag, D.: Learning low-dimensional representations of medical concepts. AMIA Jt. Summits Transl. Sci. Proc. **2016**, 41 (2016)
10. cms.gov: CMS 2008–2010 data entrepreneurs' synthetic public use file (2015)

11. Davis, J.V., Kulis, B., Jain, P., Sra, S., Dhillon, I.S.: Information-theoretic metric learning. In: ICML, pp. 209–216. ACM (2007)
12. Goldberger, J., Hinton, G.E., Roweis, S.T., Salakhutdinov, R.R.: Neighbourhood components analysis. In: NIPS, pp. 513–520 (2005)
13. Hinton, G.E., Roweis, S.T.: Stochastic neighbor embedding. In: NIPS, pp. 857–864 (2003)
14. Hochreiter, S., Schmidhuber, J.: Long short-term memory. Neural Comput. **9**(8), 1735–1780 (1997)
15. Jensen, P.B., Jensen, L.J., Brunak, S.: Mining electronic health records: towards better research applications and clinical care. Nat. Rev. Gen. **13**(6), 395 (2012)
16. Johnson, A.E., et al.: MIMIC-III, a freely accessible critical care database. Sci. Data **3**, 160035 (2016)
17. Jolliffe, I.: Principal component analysis for special types of data. In: Jolliffe, I. (ed.) Principal Component Analysis, pp. 338–372. Springer, New York (2002). https://doi.org/10.1007/0-387-22440-8_13
18. Lipton, Z.C., Kale, D.C., Elkan, C., Wetzel, R.: Learning to diagnose with LSTM recurrent neural networks. arXiv preprint arXiv:1511.03677 (2015)
19. Meystre, S., et al.: Clinical data reuse or secondary use: current status and potential future progress (2017)
20. Mikolov, T., et al.: Distributed representations of words and phrases and their compositionality. In: NIPS, pp. 3111–3119 (2013)
21. Mikolov, T., et al.: Efficient estimation of word representations in vector space. arXiv:1301.3781 (2013)
22. Miotto, R., Li, L., Kidd, B.A., Dudley, J.T.: Deep patient: an unsupervised representation to predict the future of patients from the electronic health records. Sci. Rep. **6**, 26094 (2016)
23. Peng, X., Long, G., Pan, S., Jiang, J., Niu, Z.: Attentive dual embedding for understanding medical concepts in electronic health records. In: IJCNN, pp. 1–8 (2019)
24. Peng, X., Long, G., Shen, T., Wang, S., Jiang, J., Blumenstein, M.: Temporal self-attention network for medical concept embedding. arXiv preprint arXiv:1909.06886 (2019)
25. Schoen, C., Osborn, R., Doty, M.M., Squires, D., Peugh, J., Applebaum, S.: A survey of primary care physicians in eleven countries, 2009: perspectives on care, costs, and experiences: doctors say problems exist across all eleven countries, although some nations are doing a better job than others. Health Aff. **28**(Suppl1), w1171–w1183 (2009)
26. Suo, Q., et al.: Deep patient similarity learning for personalized healthcare. IEEE T NANOBIOSCI **17**(3), 219–227 (2018)
27. Wang, F.: Semisupervised metric learning by maximizing constraint margin. Cybernetics **41**(4), 931–939 (2011)
28. Wang, F., Sun, J.: PSF: a unified patient similarity evaluation framework through metric learning with weak supervision. IEEE J. Biomed. Health Inform. **19**(3), 1053–1060 (2015)
29. Wang, F., Sun, J., Hu, J., Ebadollahi, S.: iMet: interactive metric learning in healthcare applications. In: SDM, pp. 944–955. SIAM (2011)
30. Weinberger, K.Q., Saul, L.K.: Distance metric learning for large margin nearest neighbor classification. JMLR **10**(Feb), 207–244 (2009)
31. Weiskopf, N.G., et al.: Defining and measuring completeness of electronic health records for secondary use. J. Biomed. Inform. **46**(5), 830–836 (2013)

32. Weiskopf, N.G., Weng, C.: Methods and dimensions of electronic health record data quality assessment: enabling reuse for clinical research. JAMIA **20**(1), 144–151 (2013)

33. Xing, E.P., Jordan, M.I., Russell, S.J., Ng, A.Y.: Distance metric learning with application to clustering with side-information. In: NIPS, pp. 521–528 (2003)

Data Replication Optimization
Using Simulated Annealing

Chee Keong Wee[(✉)] and Richi Nayak[(✉)]

School of Electrical Engineering and Computer Science,
Science and Engineering Faculty, Queensland University of Technology,
Brisbane, QLD, Australia
ckwee@outlook.com, r.nayak@qut.edu.au

Abstract. Data replication is ubiquitous in a large organization where multiple
IT systems need to share information for their operation. This function is usually
fulfilled by an enterprise replicating software that is dependent on the configuration that the IT administrator sets. The setup specifies the tables and routes,
but it may not be optimum to meet the workload, leading to replication's lag and
bottlenecks. This paper proposes an approach to solving the configuration
optimization problem for the data replication software with the simulated-
annealing based heuristic. Empirical results show that the configuration setting
enables the replicating software to perform at least 5 times better than the
baseline configuration set achieved by this approach.

1 Introduction

Data replication is an essential business need where data is copied to other systems to
maintain high availability, data reporting, business consolidation, workload sharing as
well as support disaster recovery standby nodes with redundancy in data sources [1]. In
order to achieve high volume and speed transfer of data, productivity tools such as data
integration and replication software are commonly used to serve the need of copying
data from sources and transform them before applying them into destination systems
[2]. IT administrators must face the challenge of managing the data replicating environment a mist the constantly growing volume of data on enterprise IT systems.

Prior research has shown that delays in the data replication occur frequently due to
software's performance constraints and thus creating bottlenecks that inhibit the software overall efficiency [3, 4]. some of the factors that affect the process are [5]:
(1) tables' attributes do limit the number of parallelized query processes that can be
made; (2) a large number of tables that needs to be involved in the replication; (3) the
volume of the data that changed among the tables; (4) the different type of data that
these participating tables' columns have; and (5) the velocity, variety, and volume of
SQL statements applied to tables involved in replications. A common approach is to
use an optimizing technique to control the parameters and configuration of the data
replication's setup [3, 4]. For example, to effectively share the workload, the transfer
rate can be manipulated by controlling the configuration of data routing queues and
tables [3]. A queue represents the channel in which the data is extracted and passed
from the source to the destination. If the requirement is of transferring the high volume

© Springer Nature Singapore Pte Ltd. 2019
T. D. Le et al. (Eds.): AusDM 2019, CCIS 1127, pp. 222–234, 2019.
https://doi.org/10.1007/978-981-15-1699-3_18

of data, all the changes that passed to the queue for transfer will build up. This may create a bottleneck if the transfer queue does not process the replication fast enough to clear the backlog. It is not desirable to have a series of tables that have a different level of low and high post activities congregating around specific queues. While some queues can clear tables with low activities and then go into idle mode, others will have a lot of highly active tables and create huge backlogs, hindering the overall replication [1, 3]. But an imbalance queue arrangement will delay the data transfer that creates an incomplete point-in-time view of the source database at the destination [2].

A possible solution is to create more queues to support the replication; however, there are several shortcomings with this approach [3]. Each queue requires additional system resources to run. The more queues that the replicating setup has, the more resource is required to support them as well an overhead will be incurred. It is desirable that replicating tables with a range of low to high changing activities should be arranged across a predefined number of queues so that the transfer of data changes for a given volume can be cleared at the least time.

We propose a novel method to optimize the configuration for the data replicating setup based on a Simulated Annealing (SA) algorithm [6]. There is no prior research made in optimizing the performance of data replications in the DB environment to our best knowledge. The computational search space to find the best arrangement of queues with tables of varying workloads is high and it is difficult to work through all the different combinations. Techniques like gradient descent and statistical methods will take a long time to solve this problem. we want to find the near-best solution within a short time limit. SA is metaheuristic and it can approximate global optimization for the data replication problem since there is a large search space. The outcome is a configuration of the best arrangement of tables with different levels of activities against a pre-defined number of queues.

Each iteration of the test was made with the activation of a unique Shareplex's configuration arrangement, followed by running a procedure that performs a variety of SQL changes on the tables at the source site. The overall duration for the entire SQL batch update to be transmitted across from the source to the target site using that configuration set was measured. Each unique set of configurations enables Shareplex to work differently, and with the SA algorithm to test and search through a list of different configurations. Empirical results show that a near-global optimum configuration was found to satisfy the DB replication's data transfer need. To our best knowledge, this is the first method based on the AI technique to optimize the performance of the data replicating tool like Shareplex.

2 Related Works and Background

Data replication serves different needs in IT organizations such as data distribution, workload sharing, reporting, backup or disaster recovery [1, 7]. For data distribution, the key requirement is that data must be shared quickly across a multitude of systems with minimum delays. For workload sharing, the key requirement is to increase a system's throughput by distributing its operation across a multitude of hardware to ease the workload. In all the replication cases, the data must be reconciled at near real-time to protect the data integrity and quality, so the changes in the data must be reflected at the repository which resides in a different location.

A survey paper [8] presented a series of time and space-based strategies used in data replication deployment, taking into consideration the characteristics and requirements of the data and systems. The survey covered the consideration of selecting the appropriate files for replication, emphasizing important and most accessed groups over those with low demands. They also proposed performance evaluation metrics to evaluate each of the different strategy's effectiveness [8]. Recently, a new evaluation metric was proposed for measuring the execution time and bandwidth consumption of data replication leading towards optimization [9]. Authors in [10] proposed the centralized dynamic scheduling and replica placement strategies that manage the data and task scheduling for optimum cost and data transfer time, in order to improve the file access time by applications across a geographically-wide data grid.

For solving optimization problems such as this, meta-heuristics is one of the more popular approaches. Meta-heuristics can be classified into two groups; population-based and single solution-based [11]. The problems that they handle are either continuous or discrete, and a problem such as finding the best queue-table arrangement for Shareplex can be considered as a combinatorial optimization problem (COP) which is regarded as NP-hard where no optimum configuration for Shareplex can be found within an acceptable time and resources [11]. The possible optimum configuration is considered as a single solution that can serve the data replication environment, so the requirement is to use single solution-based algorithms such as hill climbing or simulated annealing [11]. Other meta-heuristics such as Genetic algorithm, Ant colony or particle Swam optimization is more suitable for problems that have a population of solutions [11].

On the other hand, the Simulated Annealing (SA) method has been commonly used to solve the NP-hard problems that range from transportation and energy production to system optimization. An example is a truck-trailer routing problem where each delivery route requires to serve several customers over a defined geographical region and it needs to adhere to a constraint of allocated window periods for the delivery trucks to fulfill [12]. Another example is of finding the ideal location for the turbines to be installed across the water distribution network that can capitalize on the water flow to generate hydroelectricity [13]. SA has also been applied to solve a Multi-objective Redundancy allocation problem, improving the reliability of a system which comprised of subsystems of numerous components that each have different constraints [14].

To our best knowledge, there are no articles that are devoted to the data replication optimization problem between databases with reference on specific application tools such as Shareplex [15] or GoldenGate [4]. This research work to address the optimization of the data replication tool's configuration sets precedence.

2.1 Shareplex Data Replication

There exist several types of data replication software specific to the type of database that it supports and the functionality that the users want. While Extract-Transform-Load (ETL) software such as SAP Data Services or IBM's DataStage can be used for replicating data, they don't have the capability to do it near real-time. That is where a specialized tool such as Shareplex [3], Oracle's GoldenGate [4] and SAP's Smart Data Integration comes in [16]. Shareplex is used for this research as it is the most popular data replicating tool developed by Quest software for both commercial and open-source databases [17]. Shareplex runs in the background and performs data capture by reading the database's log files for changes to tables that are listed in its replication configuration, then propagates and applies the changes over to the target systems in near real-time. The Shareplex's framework comprises several components of data capture, read, export, import, and post, in addition to the source and target databases that they run against. Figure 1 shows how the changes are captured and transported from the source to the target databases.

The major components and subsystems of Shareplex are as followed; at the source DB, there are three main processes; (1) the Capture process reads the redo logs and archived logs for changes, then sends the change to the capture's queue [18], (2) Read process reads the data from the capture queues and processes it by repackaging them with information for network transmission. The processed data is then stored in the export queue and (3) Export process reads the processed replicated data from the export queue and transfers to the target across the network. At the target site, there are two processes; (1) Import process intercepts all transported replicated data sent out by the export process and stores them in the import queue and (2) Post process transforms the data read from the import queue into the relevant SQL statement before they can be executed against the target database. The data transport among the process at both the source and target sites are serviced by Queues. All the queues are dynamic data repositories that hold the temporary data for the duration of data capture, transmission, and reception through the process of data replication.

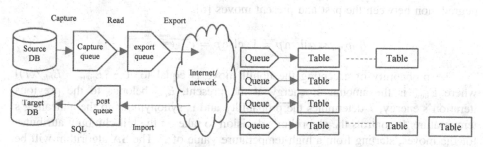

Fig. 1. Shareplex's data replication architecture

Fig. 2. Shareplex's replication configuration file structure

The replication can be set up or controlled by a configuration file as shown in Fig. 2. It defines the list of tables that need to participate in the data replication; on the source side, the schema and table's names, on the target side, the schema and object name, followed by routing information in the format (target_system:named_queue @o. Target_oracle_sid). Shareplex can support a variety of replicating architecture; from a single, bi-direction to star topology like multiple-target and multiple-source [19].

For the Shareplex's service operation, it runs under a specific UNIX user account that shares the same admin group as the Oracle database group which was used to install the Oracle binaries. There are several UNIX's environment shell parameters; $ORACLE_SID, $ORACLE_HOME, $SP_COP_TPORT, $SP_COP_UPORT, $SP_SYS_VARDIR and $SP_SYS_HOST_NAME. Once Shareplex is installed on both the source and target system, the administrator will activate a configuration file to initiate the data replication. All information, including debug and errors, are captured and stored into event logs under $SP_SYS_VARDIR/logs directory. In the event should there be tables that are out of sync, Shareplex has a compare/repair feature that allows the administrator to fix the replication tables and bring them back into synchronization [20].

2.2 Simulated Annealing

SA is an analogous method for optimization that attempts to find the global optima among the large landscape of local optima of solutions for a problem environment [21]. Annealing is a process where metals are heated and then cool down to a hardened condition. The objective function represents the energy of the material. SA has a similarity with a hill-climbing algorithm [21] with the exception that it doesn't just pick the better move in its iteration but rather a random one [6]. Referring to Eq. (1), if the selected move can improve the solution, it will be accepted. However, this single goal can cause the algorithm to be stuck in local optima. Hence, the algorithm also takes a chance in making a choice to accept a worse move based on some probabilities of a value that is less than 1. It starts with a high probability which means that the algorithm will be more liberal to accept the bad move but that will decrease rapidly with the degradation between the past and present moves [6].

$$If\ q_{limit} < q \text{ then } p = 1, \text{ else } p = e^{\frac{cost\_new - cost\_old}{temperature}} \tag{1}$$

The probability of accepting the uphill move is equal to, $1 - (E_{new} - E_{old}/kT))$ where E_{new} is the amount of energy at the present, E_{old} belongs to the previous iteration's energy, T determines the probability and is synonymous to the annealing's temperature. It controls the algorithm's decision to take on the hill-climbing attributes for the moves, starting from a high-temperature value of T. The SA algorithm will be open to accepting the hill-climbing process. But as T decreases over time in energy, this probability will decrease, and the algorithm will be less inclined to accept this until T reaches zero. k is the constant that relates the temperature to the energy [13]. SA is commonly used to solve the optimization problem in large and discreet configuration spaces that can be considered as nondeterministic NP-complete problems such as travel salesman problem where the combination space of queues to tables arrangement is large and an approximating global solution is required with a specific period of computational time [6].

3 The Problem Definition

Contemporary data replicating software runs with processes and queues [15]. They function by constantly reading the source databases for changes. The captured data are then copied and propagate to the designated databases and are applied with minimum delays. Ideally, these changes should occur with minimum delays. However, the SQL change activities are IO intensive on the databases and will incur waits and latency on the application of the changes. This delay will increase substantially if there are huge volumes of changes that need to be replicated across, causing long delays and creating backlogs that impact on the overall replication performance [1].

In order to alleviate this challenge, prior research recommends having multiple queues running in parallel so as to spread out the entire job by breaking it into smaller sections among the queue [3]. Usually, the highly active tables will be assigned to one set of queues and other slower ones are bundled together under other queues [5]. However, there are a finite number of queues that can be made available as more queues will equate higher system overheads and greater IT administration effort. In summary, we do not want to create thousands of queues to support the replications nor do we want to create too few queues that they are not sufficient to support the data replication effectively [2]. Another factor is that we do not know which high or low active tables can give the best combination for the replication process that can create the least amount of backlog clearing time.

The combination of tables of varying DML activities across a series of queues can be regarded as a nondeterministic polynomial time (NP)-hard problem in combinatorial optimization [5]. The objective of the research problem of data partition is to determine the tables-queue allocation plan with the timing required to clear a series of tables that will receive a different level of changes to simulate their activity in the databases. However, there exist complications that will impact the calculation. They are: (1) Tables don't have a constant pattern in change activities. (2) There is a large variability in tables' properties, activities and characteristics in different databases [22]. (3) The IT environment's capacity and performance that DRE runs on, including the OS, network, DB. [1]. We must balance the load between the tables' activities across the queues in such a way that the queues can transfer all their changes across with minimum delays and lags. We cannot treat this as a linear problem as the activities on the tables and volumes are neither linear nor consistent throughout the IT system operation [23].

4 The Proposed SA Based Approach

We propose to solve the queue configuration optimization using a simulated annealing algorithm as shown in Fig. 3. For this experiment, Shareplex is used as the replicating tool. Shareplex is configured to operate at both source and target DBs, with a series of tables setup to replicate between them. The mode of the replication transfer process is controlled by the Shareplex configuration file at the source DB. This file has direct impact on the performance of the overall replicating process. Referring to Table 1, one queue can service several tables, and, in some cases, it is able to support all of them if the number of tables is small and their DML activities are low. However, when the replicating system is large and their activities increase exponentially, the load will flood the queue. The queue will not have enough capacity to handle them and this will form the bottleneck in the replication process, causing massive build-up [3].

The goal is to distribute the tables across to a series of queues to spread out the load. Ideally, if all the tables have similar data contents with an equal volume of DML activities, then the load can be easily split evenly across a given number of queues. In the real work environment, the tables in an IT system will have a wide variation in terms of datatypes, contents, volumes and work activities. Each table has its own characteristics that impose a level of impact on the replication's queues. Bundling them up randomly in queues without due consideration will not only create unnecessary segregation of loads on certain queues but affect the overall replication throughput. Some of the queues with highly active and bigger load tables will take much longer for the backlogs to clear as compared to the other tables that have low volume and activities. The challenge here is to approximate. For example, the global optimum arrangement of tables with different load characteristics to be serviced by a given number of queues so that the data replication process can clear the backlog in the replication in the least amount of time. Another method of solving an NP-hard problem is to try all the combinations available but that will be very expensive in terms of time and computational cost.

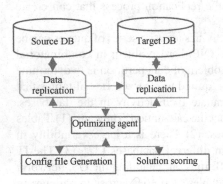

Fig. 3. Optimization of Shareplex's data replication

Table 1. Shareplex's configuration file Datasource:o.SOURCE_DB

#Source table	Destination table	Routing map
HR.COUNTRIES	HR.COUNTRIES	src_svr:Q0*tgt_svr@o.
HR.DEPARTMENTS	HR.DEPARTMENTS	tgt_db
HR.EMPLOYEES	HR.EMPLOYEES	src_svr:Q0*tgt_svr@o.
HR.JOBS	HR.JOBS	tgt_db
HR.JOB_HISTORY	HR.JOB_HISTORY	src_svr:Q1*tgt_svr@o.
HR.LOCATIONS	HR.LOCATIONS	tgt_db
HR.REGIONS	HR.REGIONS	src_svr:Q1*tgt_svr@o.
		tgt_db
		src_svr:Q2*tgt_svr@o.
		tgt_db
		src_svr:Q2*tgt_svr@o.
		tgt_db
		src_svr:Q2*tgt_svr@o.
		tgt_db

4.1 Estimating the Cost of the Solution

The solution in this paper is a configuration file that comprises of queues with tables allocated to them. The cost of a configuration-solution cannot be represented as a formula but through a series of application's process against the Shareplex's data replicating environment. the optimum configuration may not be able to yield the best throughput as the workload in any IT system's databases in the real world tends to fluctuate from time to time. It is also difficult to predict the ability of the replication as there are many dynamic factors involved in the environment setup [3]. Our estimation attempt takes the following into consideration; (1) The list of tables and their data change activities for a defined period. (2) The allowable number of queues. An IT administrator can specify the max number of queues that can handle. Any excess number of queues above this threshold will be penalized but that shouldn't stop the algorithm from considering them. If the additional queue(s) above the preference can bring in more benefits in comparison to the penalty it suffered, then it should be considered. (3) The measurement for a single queue performance will span from the source to the target databases. The weakness of a complete stack of replication's setup lies with the weakness queues that are the most susceptible to experience backlogs the most.

The replicating process in a given queue begins at the source database; starting from the Data Capture process, then to the Read and followed by an Export process. The Export process transfer the information to the target DB's Import process, which in turn divert to the Post process that converts the information into SQL statements and applies to the target DB. Each queue holds the information's backlogs and has two statistics to show its progress activity: (1) the number of data statement changes; and (2) the time required to clear them. In the proposed approach, we sum up the statistics of each queue that determines its capability to handle the replicating load. The approach is to optimize the arrangement of the replicating tables across the most preferred number of queues, therefore, the cost of the configuration setting solution can be deemed as the summation of each queue that contains n numbers of tables, each with its own number of rows, row length, number of SQL updates and the time period that the load occurs. A cost factor is introduced to instill cost of exceeding the allowable number of queues allowed as in Eq (2);

Algorithm 1. SA for finding optimum Shareplex's configuration

```
Input: Shareplex configuration solution S, Length L
initialize1: Acceptance_probabilty A, random_number r, neighbouring solution of S is S'
Initialize2: temp=1000, cooling constant=0.9, temp_min=1, old_cost=0
Initialize3: old solution cost cost_old, new_solution_cost cost_new
Result: approxmiate global optimum solution, S_opt
While temperature > temp_min do
  For i = 1 to L do
    #procedure to generate new configuration file
    S' = sp_generate_new_config
    cost_old = cost(S)
    cost_new = cost(S')
    if (cost_new - cost_old) <=0 then
      S = S'
    end
    if (cost_new - cost_old) >=0 then
         cost_new-cost_old
      A = e   temperature
      r = random(1)
      #test probability to accept new solution
      If A > r then
        S = S'
        cost_old= cost_new
      end
    end
    T= rT    #Cooling temperature
    #procedure to restore all tables at source_site
    sp_restore_source_tables
    #procedure to resync all tables at target site to source
    sp_resync_target_tables
end
s_opt=S
```

$$\text{solution} = (p * \frac{1}{q} \sum_{n=1}^{m} (n_h * n_k * n_q))_t \tag{2}$$

Where n refers to the replicating tables, h is the number of rows that n has, j is the number of SQL workload activities are performed against the table, k is the size of a single row in the table, t is the time period, q is the number of queues and p is the cost of the queue maintenance. If it is less than the allowable threshold, the cost is 1. But if it exceeds the threshold, then the cost will be much higher as followed. This is to prevent the method from allocating an excessive number of queues that are beyond the allowable range as specified by the IT administrator. The proposed approach's algorithm is listed in Algorithm 1.

5 Empirical Analysis

The experiments were conducted using two virtual machines that run on Linux OS and support two Oracle databases with Shareplex configured to replicate tables from a source DB to another. Each virtual machine has 4 GB of memory. The DB tables belong to a common DB schema and they comprised of a variety of data types; numeric, integer and varchar. We have created procedures that simulate a series of DML activities of these tables. The activities can be segregated into three groups; low, medium and high. Tables with low change activities receive less than 100 DML statements whereas those with high activities will get > 10000. The DML activities comprised of a mixture of delete, insert and update statements, all of which will be made on the tables at the source DB. The DML changes will then be propagated and applied to the target DB based on the Shareplex's configuration file. The test setup is a controlled environment with no other IT application running against them. The maximum number of queues allowed is 5 for this test. 10 tables have been set up to replicate from the Source to the Target DB and a load test procedure was written to simulate the DML activity against the 10 tables with 100, 2500, 5000, 10000, 12500, 15000, 17500, 20000, 30000 and 40000 iterations.

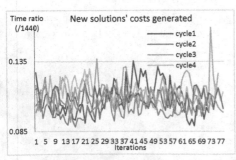

Fig. 4. Generated random solutions' costs

Fig. 5. SA's selected solutions' costs

For the simulated annealing algorithm, the control parameter is set with the bigger starting temperature and a smaller cooling which have a better chance of finding the optimal solution but require more iteration and time to execute. The final control parameters are used are as followed: starting temperature = 1000; final temperature = 1; and cooling rate – 0.9. Random solutions are generated throughout the process in every iteration under each temperate diminishing cycle. The randomly generated solutions' costs are tracked in Fig. 4 and are measured in seconds. Their values are relative to the number of minutes in a day which is 1440.

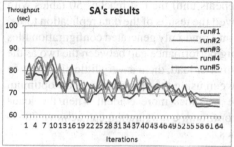 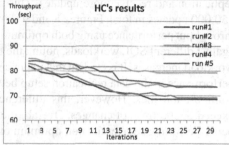

Fig. 6. Simulated Annealing (SA) results   **Fig. 7.** Hill Climbing (HC) results

Figure 5 showed the initial randomness in the selection of the solutions based on the SA' algorithm in cycle 1. It started to converge at the 73rd iteration in cycle 1 onward. From the 2nd cycle onward, the results remain the same, adhering to the optimum solution with a cost of 0.085. This test has been repeated and the pattern of converging is the same which occurs during the 1st cycle while the other plateaued at the discovered optimum solution. Figure 6 showed the results from the five trial runs with the first cycle result shown. While the optimum configuration is the same dis- covered from the five test runs, their completion time has some slight variation, and this could be impacted by the OS environment against the Virtual Machines that the experiment is running on.

For the five test runs, the constraints set are on the number of iterations and queues available to support the table replication. As shown in Fig. 6, at the beginning where all the replicating tables are assigned to a single queue, the observed cost for the solution was above 80+ value. as the iteration proceeds, the proposed approach accepts solu- tions with a higher cost than the initials which is expected. As the temperature energy starts to cool down in SA, the tolerance level and probability threshold to accept worse off solutions is getting lower. During the initial half of the iterations, a higher level of fluctuation is seen among cost solutions that were accepted. But toward the end of the run, the solutions with the lower cost have less volatility and remain around 63+ values. In Table 2, the results that were derived using the simulated annealing and hill- climbing heuristics. Compared with Fig. 7 where the tests are conducted with a similar setup but with the Hill Climbing heuristics [24], the latter method only accepts better solutions' costs and converges to solutions that are deemed as local optima. The results discovered by both heuristics can't converge to the same cost for the different test runs.

The reason is that they can approximate the optimums and the test setup could have minor fluctuations workloads and operating system activities that could influence the overall controlled environment.

The next test is to compare the difference in the data replications' throughput between the optimum configuration file that is discovered by the SA method from the various test runs, versus another one that has been randomly generated. We subject the test under a variety of DB workload which runs a series of SQL updates to the tables and observes the time taken to clear the replicating backlogs, starting with 10 tables in replications for the 1st run. The subsequent test runs will have an increment of 10 more participating tables in replication and receive more updates. This repeats until the test reaches 50 tables with 550,000 updates in total. Figures 8 and 9 showed the results of the data replication tools' throughput performance using both optimum and randomly generated configuration files against different SQL workloads, noting the big performance gap between the two as the workload increases. It is evident that under a smaller load, there is very little difference in performance in the data replication setup between using the optimum and non-optimum configuration files. However, this difference becomes more evident when the loads exceed 55000 rows of changes. Overall, the optimum configuration can achieve 20+% better performance over the non-optimum ones

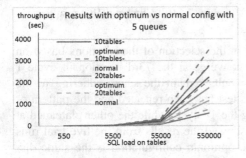
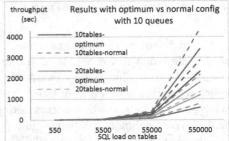

Fig. 8. Performance results using optimum and non-optimum configuration setting for 5 queues

Fig. 9. Performance results using optimum and non-optimum configuration setting for 10 queues

Table 2. Throughput results from optimum configuration found by SA and HC heuristics

Runs	Optimum solution throughputs (sec) by SA	Number of queues derived by SA	Optimum solution throughputs (sec) by HC	Number of queues derived by HC	Tables' workload
1	**64.26**	5	**68.6**	5	1000
2	65.29	5	69.1	5	1000
3	66.16	5	73.6	5	1000
4	67.46	5	79.3	5	1000
5	69.36	5	73.6	5	1000
Average	66.50	5	72.84	5	1000

6 Conclusion

A simulated annealing-based method has been proposed to find the optimum arrangement of tables for a pre-determined number of queues for the data replication setup that consists of databases and data replicating software. While the software used in our setup is primarily Oracle databases and Shareplex, the concept of using the algorithm to find the optimum configuration for the data replication software is generic and applicable to others. The simulation results demonstrate the effectiveness and efficiency of the proposed SA method. In the current context of data replication, the method to improve their throughput has been largely based on a series of vendors' best practices and IT administrators' experiences.

References

1. King, E.: Automated Database Refresh in Very Large and Highly Replicated Environments (2011)
2. Simitsis, A., Vassiliadis, P., Sellis, T.: Optimizing ETL processes in data warehouses. IEEE (2005)
3. Quest Software: Shareplex 9.0 Reference Guide (2018)
4. Gupta, R.: Introduction to Oracle GoldenGate (OGG). In: Gupta, R. (ed.) Mastering Oracle GoldenGate, pp. 3–10. Springer, Heidelberg (2016). https://doi.org/10.1007/978-1-4842-2301-7_1
5. Gill, N.K., Singh, S.: A dynamic, cost-aware, optimized data replication strategy for heterogeneous cloud data centers. Future Gener. Comput. Syst. **65**, 10–32 (2016)
6. Chopard, B., Tomassini, M.: Simulated annealing. In: Chopard, B., Tomassini, M. (eds.) An Introduction to Metaheuristics for Optimization, pp. 59–79. Springer, Cham (2018). https://doi.org/10.1007/978-3-319-93073-2_4
7. Quest Software: SharePlex 9.0 - Reference Guide. Quest Support (2018)
8. Souravlas, S., Sifaleras, A.: Trends in data replication strategies: a survey. Int. J. Parallel Emergent Distrib. Syst. **34**, 1–18 (2017)
9. Hamdeni, C., Hamrouni, T., Charrada, F.B.: Evaluation of site availability exploitation towards performance optimization in data grids. Clust. Comput. **21**(4), 1967–1980 (2018)
10. Nazir, B., et al.: The impact of the implementation cost of replication in data grid job scheduling. Math. Comput. Appl. **23**(2), 28 (2018)
11. Blum, C., Roli, A.: Metaheuristics in combinatorial optimization: Overview and conceptual comparison. ACM Comput. Surv. (CSUR) **35**(3), 268–308 (2003)
12. Assadi, M.T., Bagheri, M.: Differential evolution and Population-based simulated annealing for truck scheduling problem in multiple door cross-docking systems. Comput. Ind. Eng. **96**, 149–161 (2016)
13. Samora, I., et al.: Simulated annealing in optimization of energy production in a water supply network. Water Resour. Manag. **30**(4), 1533–1547 (2016)
14. Zaretalab, A., et al.: A knowledge-based archive multi-objective simulated annealing algorithm to optimize series–parallel system with choice of redundancy strategies. Comput. Ind. Eng. **80**, 33–44 (2015)
15. Connell, A.M.: An analysis of database replication technologies with regard to Deep Space Network application requirements (2011)
16. Bahl, A.: Use Case to S/4HANA Smart Data Integration (SDI), SAP Editor (2018)

17. Quest Software: Shareplex for Oracle v9.1.4 (2018)
18. Brunt, B.: Going for gold: Dell Software's SharePlex database replication offering is a powerful tool with a small footprint. Computer Reseller News (UK), p. 23 (2016)
19. Dell Software Extends SharePlex to Optimize Data Integration and Analysis. Information Technology Newsweekly, p. 136 (2013)
20. Quest Software Releases SharePlex v9, in ICT Monitor Worldwide U6 - ctx_ver=Z39.88-2004&ctx_enc=info%3Aofi%2Fenc%3AUTF-8&rfr_id=info%3Asid%2Fsummon.serialssolutions.com&rft_val_fmt=info%3Aofi%2Ffmt%3Akev%3Amtx%3Ajournal&rft.genre=article&rft.atitle=Quest+Software+Releases+SharePlex+v9&rft.jtitle=ICT+Monitor+Worldwide&rft.date=2017-06-22&rft.pub=SyndiGate+Media+Inc&paramdict=en-US U7 - Newspaper Article, SyndiGate Media Inc., Amman (2017)
21. Nikolaev, A.G., Jacobson, S.H.: Simulated annealing. In: Gendreau, M., Potvin, J.Y. (eds.) Handbook of Metaheuristics, vol. 146, pp. 1–39. Springer, Boston (2010)
22. Quest Software: SharePlex 9.0 - Administration Guide (2019)
23. Milani, B.A., Navimipour, N.J.: A comprehensive review of the data replication techniques in the cloud environments: Major trends and future directions. J. Netw. Comput. Appl. **64**, 229–238 (2016)
24. Al-Betar, M.A.: β-Hill climbing: an exploratory local search. Neural Comput. Appl. **28**(1), 153–168 (2017)

Readiness of Smartphones for Data Collection and Data Mining with an Example Application in Mental Health

Darren Yates$^{(\boxtimes)}$ and Md. Zahidul Islam

School of Computing and Mathematics, Charles Sturt University,
Panorama Avenue, Bathurst, NSW 2795, Australia
{dyates,zislam}@csu.edu.au

Abstract. Smartphones have become the ultimate 'personal' computers with sufficient processing power and storage to perform machine learning tasks. Building on our previous research, this paper investigates an example practical application of this capability, combining it with a smartphone's on-board sensors to develop a personalised, self-contained machine-learning framework for monitoring mental health. We present a mobile application for Android devices called 'Mindful' that incorporates data collection from the phone's sensors and data sources, pre-processes the data locally and executes data mining on that data to provide pre-emptive feedback to the phone user about their mind state. Rather than as a finished product, this application is presented as a first step to show that from a technological perspective, smartphones are well equipped to perform this type of role. We invite colleagues from the mental health sciences to join us in furthering this work into a smart monitor for mental health.

Keywords: Smartphones · Data mining · Application · Mental health

1 Introduction

Since their introduction, smartphones have revolutionised the way we communicate. Continued improvement in computing performance has combined with declining hardware costs, particularly in the entry-level market, to create smartphones of very low cost yet with considerable processing power. In Australia, 'pre-paid' phones (without contract but locked to a network provider) regularly sell for under $40. Even in this entry-level market segment, smartphone technology has transformed mobile computing, offering capabilities approaching those of laptop computers.

In a previous paper [1], we established that smartphones are ready to perform local data mining from the perspective of processing speed and accuracy, as well as battery life and thermal/heat-dissipation considerations. Moreover, in a second paper [2], we further established that a general-purpose data mining application for smartphones could support loading a dataset and executing one of a range of common algorithms for building and evaluating models. That application is today called 'DataLearner' and is available on the Google Play store[1] for any Android 4.4 or later phone or tablet.

---

[1] https://play.google.com/store/apps/details?id=au.com.darrenyates.datalearner.

© Springer Nature Singapore Pte Ltd. 2019
T. D. Le et al. (Eds.): AusDM 2019, CCIS 1127, pp. 235–246, 2019.
https://doi.org/10.1007/978-981-15-1699-3_19

However, neither research effort involved data collection. To fill that void, this paper presents an example framework and practical implementation of mobile data collection and data mining with application in mental health.

With the global population of depressed persons estimated to have exceeded 300 million in 2018 [3], mental illness continues to be a significant issue facing the administration of public health care around the world. In Australia, nearly half (45%) of all adults between the ages of 16 and 85 will endure a mental health condition at some point in their lives [4]. The ethical issues of using smartphones within a mental health treatment framework are beyond the scope of this paper. Instead, this paper is a first step showing the technological capabilities of smartphones for data mining with reference to this specific application. Moreover, we invite our colleagues in the mental health sciences to join with us in combining data collection and data mining with today's smartphones to develop a smart mobile mental health monitoring system.

1.1 Original Contribution

The original research contributions of this paper include:

- Development of a time-focused data collection and preprocessing framework that builds into a dataset suitable for mobile data mining (See Sects. 3.1 and 3.2)
- Embedding of data mining methods within Android devices (Sects. 3.3 and 3.4)
- Implementation of this research as a stand-alone Android application called 'Mindful' (Sect. 4). Data privacy is also preserved due to Mindful's local computation.

The remainder of this paper continues with Sect. 2 presenting a short survey of current mobile mental health research, while Sect. 3 explains our framework. Section 4 details its implementation, Sect. 5 demonstrates the framework and application on synthetic data, followed by a brief discussion and conclusion in Sect. 6.

2 Related Work

Smartphones have formed a focal point for recent mental health research [5–7]. However, in the studies reviewed, the smartphone typically performs the role of data gatherer and transfers that data to an external server. This raises potential issues in security, privacy and connectivity. Arora, Yttri and Nilsen [8] point out that security and privacy remain "an on-going concern for researchers conducting mHealth [mobile health] studies". Users rely on the security of the external server to ensure the privacy of their data is maintained, while unreliable connectivity may delay or prevent the data being transferred, possibly lowering its value to both the user and the research.

Smartphones have an application opportunity in mental health through the close relationship users have with their devices. The authors of [9] suggest users of smartphones in general are "empowered by the confidentiality of their engagement" with their devices. However, mental health apps that transfer personal data to external servers are subject to the security and privacy arrangements of those servers. The rise of fitness bands or 'fitness activity trackers', devices measuring physical activity, has seen their growing inclusion in mobile health research, including mental health research, where they measure physical activity by day and sleep patterns at night. However, these devices face similar issues to those noted for smartphones.

In their research, the authors of [10] supplied fitness bands to study participants as required. Yet, it is known that low socio-economic status correlates to higher rates of depression [11] and the purchase cost of a fitness band, if required, may pose a burden on mental health sufferers. By contrast, even those pre-paid smartphones available in Australia for under $40 include an accelerometer sensor comparable to that found in many fitness bands. Using a smartphone exclusively also makes sense from a patient perspective. According to a 2014 study, 77% of Australians experiencing homelessness (an issue known to correlate with mental illness [12]) had a smartphone [13]. This aligns with a 2018 report that 89% of Australians now own a smartphone [14]. Compared with the purchase cost of a fitness band, a free mobile app has the potential to reach a much larger patient population.

Moreover, by having data stored and analysed on the phone itself, security, privacy and connectivity issues can be largely mitigated. Previous studies have implemented smartphone features that log direct human interaction with the device, as well as gather physical activity data through fitness bands [10, 15]. However, it has been noted in [16] that while smartphone accelerometer sensing was not as accurate as fitness bands, smartphones were still able to provide useful results. Smartphones also provide the user the benefit of only having to keep track of one device. Where a research gap exists is using the phone to capture, process, analyse and respond to the user's data.

3 Our Framework

Smartphones are ideal for incorporating into mental health services, not just for the relationship users increasingly have with their devices, but also for the array of sensors available within these devices. By combining this sensor array with direct user input and local machine learning, a self-contained framework can be developed that automates sensor sampling at regular intervals, creates local data records, mines that data for knowledge and provides timely feedback to the user. A simplified block diagram of our framework is presented in Fig. 1 and incorporates the following steps:

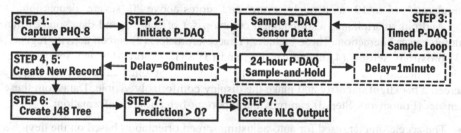

Fig. 1. A simplified block diagram of our mobile mental health machine-learning framework.

1. The user submits a PHQ-8 Patient Health Questionnaire [17] within the device
2. This initiates 'passive data acquisition' (P-DAQ) sampling of phone sensors
3. P-DAQ sampling occurs once per minute with the last 24-hours of data stored within 'sliding window' memory blocks (a 24-hour sliding 'sample and hold').
4. The PHQ-8 response also initiates a new dataset record that is added to the dataset.
5. A new record is created from P-DAQ data each hour to test the user's current state.
6. Step 4 triggers a new J48 (C4.5) decision tree to be built from the growing dataset.
7. The new record is tested against the J48 tree (Step 6). If a non-zero mind state is predicted, a natural language generation (NLG) user response is created.

Thus, the framework operates two key processes – it provides feedback each time the user enters a PHQ-8 response, but it also combines data acquisition with data mining at one-hour intervals to automatically predict if the user's current activity data matches that which the user has previously indicated led to a decline in mental health. If so, the user can be informed of ways to modify their behaviour.

This section will now detail the framework components: data acquisition (Sect. 3.1), data preprocessing (Sect. 3.2), knowledge discovery (Sect. 3.3) and user feedback (Sect. 3.4).

3.1   Data Acquisition (Framework Steps 1–3)

To directly capture the user's perceived mental state, the framework requires a standardised system. The PHQ-9 Patient Health Questionnaire [17] is a well-regarded tool for determining depressive disorder in adolescents and adults [18]. It consists of nine questions to identify the user's mental health over the past two weeks. Our framework modifies this period to just the past 24 hours, however, daily questionnaire responses are not required, as the framework supports more random entry patterns.

PHQ-8 is a revised version of PHQ-9 removing a question on suicidal ideation and is recommended for research applications where "data is being gathered in a self-administered fashion rather than by direct interview…" [17]. Further, the research authors found no significant loss in depression detection with the question's removal.

The PHQ-8 questions (Framework Step 1) are answered on a four-step scale from 'not at all' (scored as '0') to 'nearly every day' ('3'), relabeled here as 'all the time'. Scores are summed and converted to one of five category ratings – scores below four (4) equate to 'no' depression, scores 5 to 9 indicate 'mild' depression, 10 to 14 'moderate', 15 to 19 'moderately severe' and scores above 20 'severe' depression.

Android smartphones also feature a common set of sensors and data sources that monitor user interaction. These include: (1) accelerometer, (2) screen activity register, (3) proximity sensor, (4) phone call log, and (5) SMS (Short Message Service) log. Many Android phones also include (6) a light sensor to automatically vary the display screen's backlight intensity and make the display comfortably visible. Data from these sensors (Framework Step 3) can provide important behavioural information:

- The accelerometer, used for auto-adjusting screen orientation based on the device's physical position, can also detect user movement. A previous study showed significant correlation between physical activity and level of depression [19].

- The screen activity status can be used to indicate a user's phone activity patterns. Detecting the screen activity status and light readings can also provide indications of sleep patterns, a factor research has shown affects depression levels [20].
- Previous research has also revealed the link between social interaction/isolation and depression [21]. Smartphones provide a measure of social interaction through the phone call and SMS logs, such that their rate of use can be monitored.

3.2 Data Preprocessing (Framework Steps 3-5)

By combining the user's PHQ-8 response with this passively-collected sensor data over the same 24-hour period, a record can be created that correlates the user's self-evaluated PHQ-8 state with their social/physical interaction with their smartphone (Framework Step 4). These records then form a dataset that can be mined to identify not only activity patterns that may have affected a user's mental health previously, but also to predict their present mental health state at other times, with a view to warning the user of any change (Framework Step 7). Each record reflects the user's phone interaction over a 24-hour period. This involves recording data from the six phone sensors noted in Sect. 3.1 once per minute for 1,440 data-points per 24-hour period.

It is unrealistic to expect the user to fill in the questionnaire each day, let alone the same time each day. Therefore, we need to ensure the previous 24-hours' worth of passive data is always available, so whenever the user chooses to fill in the questionnaire, this data is applied to the questionnaire result to form a dataset record. This involves creating 24-hour 'sliding window' memory blocks that store the last 1,440 data-points (Framework Step 3). Each data point is a binary (0, 1) value indicating sensor activity. One memory block is required for each of the six phone sensor/log functions, plus another for time. The 'sliding window' in Fig. 2 works such that when a new set of data-points from the phone functions is recorded, the oldest data-point in each memory block is removed, ensuring 24-hours' worth of contiguous data.

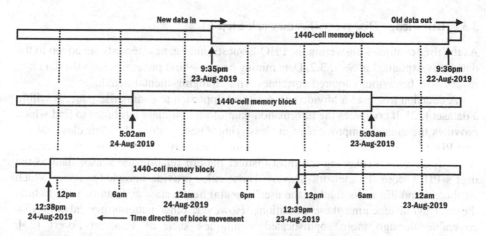

Fig. 2. The 24-hour memory block captures 1440 data points in a sliding window.

3.2.1   Creating a Dataset

To provide each dataset record with further granularity and to understand the effect time-of-day has on mental health, the six sensor memory blocks are further divided into four sub-blocks based on the time of day of each capture, shown in Fig. 3. Each sub-block covers a six-hour period, equaling 360 data-points, sampled once per minute. Data captured between midnight and 6am is designated as 'AM1', 6am to 12pm as 'AM2', 12pm to 6pm 'PM1' and 6pm to midnight 'PM2'. The individual binary scores of each data-point in a sub-block are summed to create a single value of between 0 and 360 for that block. For example, if the screen-activity memory block records 37 separate minutes of activity between midnight and 6am, the 'screen-activity-AM1' score becomes '37'. Six phone sensor/log functions, each subdivided into four six-hour time periods, creates 24 separate sub-blocks. Each sub-block becomes an attribute, giving 24 attributes. In addition, the hour of day in 24-hour time (0 to 23), the day of week (0 to 6) and the month of year (0 to 11) are captured as three additional attributes. Finally, the PHQ-8 score is added as the 'class' attribute. The result is a record comprising 28 attribute values, created and added to the dataset when the user submits their PHQ-8 questionnaire response (Framework Step 4).

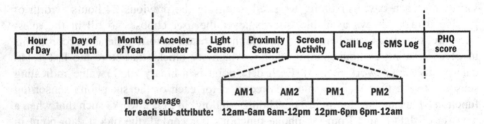

Fig. 3. The six phone features of a dataset record are divided into four six-hour timed sub-blocks - AM1, AM2, PM1 and PM2.

3.3   Knowledge Discovery (Framework Step 6)

As the user continues answering the PHQ-8 questionnaire, new records are added to the dataset as explained in Sect. 3.2. Data mining can now find patterns within the data that reveal how behaviour captured correlates with changing mental health.

A decision tree uses a 'divide-and-conquer' approach to summarise patterns within a dataset [22]. It compares the information gain of all non-class attributes to find which provides the greatest improvement in classifying dataset records by their class values. The PHQ-8 value from Sect. 3.1 is used as the class attribute. The result is a tree-like set of logic rules explaining the most distinct pattern relating the sensor data to the user's PHQ score (Framework Step 6). Those rules provide knowledge about which attributes (and life factors) affect the user's mental health state and can be used to help the user avoid repeating these conditions. However, other patterns may also be discoverable through more sophisticated techniques such as Random Forest [23], improving detection accuracy.

3.4 Present Mood Detection and User Feedback (Framework Step 7)

These logic rules can also be used to detect current mental health levels. For example, a new record can be created from current data within the 24-hour memory blocks. This record is without a PHQ-8 score, but by applying it to the logic rules of the existing decision tree, a PHQ-8 score can be predicted. Thus, a non-zero score (indicating a level of depression) can be used to warn the user that current conditions match those previously recorded as indicating mental ill-health (Framework Step 7). Moreover, Natural Language Generation (NLG) [24] can use the predicted PHQ score to create plain-text 'natural language' user feedback. Recent research into mental health app development recommends promoting activities to immediately improve the user's mood [9]. NLG provides a more personalised response than the PHQ-8 score alone.

Since this detection can be initiated at any time, how often should it be implemented? On one hand, the more often it runs, the sooner the user can be warned of any impact to their mental health. However, overly-aggressive use of detection also has drawbacks. First, setting the detection process rate too often could induce fatigue on the part of the user, resulting in either the user ignoring or obsessing over any warnings. Second, aggressive detection may also reduce the device's battery life. Given sensor sampling rates, a detection rate of once per hour provides a useful start point.

4 Implementation

Our framework has been implemented in a prototype mobile application ('app') called 'Mindful' for Android 4.4 or later devices, covering 96.2% of all Android devices in use, as of May 2019. Mindful employs four key processes – (1) learning the user's mental health state through the PHQ-8 Questionnaire; (2) automated sampling of the phone's sensors and SMS/phone logs at one-minute intervals; (3) mining the dataset for knowledge and (4) auto-detecting the user's mental health. These are explained through Sects. 4.1, 4.2, 4.3 and 4.4, while Sect. 4.5 outlines the NLG user feedback system.

4.1 Learning the User's Perceived Mental Health

Mindful learns the user's mental health state through the PHQ-8 questionnaire. This is achieved via 'slider' controls on the user interface for each question, as shown in Fig. 4a. The PHQ-8 score is summed over the eight questions and converted into the app's PHQ-8 category score attribute with a range of zero ('no' depression) to 4 ('severe').

To avoid issues of missing data, the user is assumed to have answered all questions when the app's submission button is pressed. Moreover, the button does not appear until the user scrolls through all questions. To guard against 'false positive' responses, each question's default answer is 'no depression', ensuring the user must physically answer for a non-zero depression level to be detected.

4.2 Initialising Passive Data Acquisition (P-DAQ)

Android supports the six phone sensor/log functions outlined in Sect. 3.1. Upon app installation and completion of the PHQ-8 Questionnaire, Mindful initiates the P-DAQ system to begin continuous sampling of the sensor/log data once per minute. This process continues until either the device battery is exhausted or the user turns off the data acquisition system via the user interface. The data captured is stored in integer arrays replicating the 'sliding-window memory blocks' described in Sect. 3.2. Upon a questionnaire completion, the app creates a record from the current memory block data, the three time-related attributes and PHQ-8 score, adding it to the dataset.

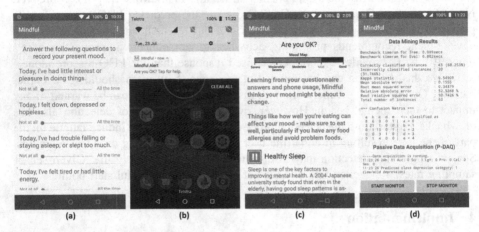

Fig. 4. The Mindful prototype app screens include (a) the PHQ-8 questionnaire, (b) notification alert, (c) an NLG-based warning alert and (d) the prototype's internal data mining results page.

4.3 Modelling the Dataset

The Mindful app looks for patterns within the phone sensor data, communications logs, and the user's previous PHQ-8 questionnaire scores that explain events leading to a change of a user's mental health state. Because the app relies on user data, the resulting model will be unique to each user. Moreover, once sufficient data records have been collected, the app can continue auto-detecting depression levels even if the user stops filling in the questionnaire, thus becoming a self-sustaining 'smart' system.

Weka is an open-source data mining suite developed by the University of Waikato for Windows, Mac OS X and Linux computers [25]. Mindful employs Weka's J48 decision tree algorithm, a Java implementation of the C4.5 algorithm developed by Ross Quinlan [22]. J48 processes the dataset, looking for patterns between the PHQ-8 ('class') attribute and sensor attributes to develop a set of logic rules. Once the user submits a new PHQ-8 response, J48 builds a new tree based on this updated data. Mindful's model of the user's life patterns is thus updated on each new record, so that detection is always performed on the latest data and follows any changing user trends.

4.4 Identifying Changes in Mental Health

While the decision tree detects patterns within the dataset upon completion of the PHQ-8 questionnaire, it can also predict depression levels on new memory block data at any time. For this, Mindful creates a new 'unlabelled' record (one without a PHQ-8 'class' score) from the 24-hour memory block data each hour and tests it against the current tree model. If the class value predicted represents a non-zero depression score, the app immediately notifies the user through Android's 'notification alert' system. This appears in the notification area (Fig. 4b). Tapping this alert opens the Mindful Alert page (Fig. 4c). This proactive feedback alerts the user that their mental health state may change and offers help to take action before-hand.

4.5 User Feedback

Mindful provides two opportunities for user feedback – following submission of a PHQ-8 response, and again after a notification alert. On each occasion, the user receives three forms of feedback, as shown in Fig. 4c. First, Mindful creates a 'mood map', a fuel gauge-like display showing the current PHQ-8 category scale of the user's mood. On answering the questionnaire, the mood map indicates the PHQ-8 category. When triggered via notification alert, it displays the predicted depression category based on current data via the decision tree model. This is followed by a natural language response aligned with the mood map that encourages the user to make life-style changes to improve wellbeing. This uses an NLG system as described in Sect. 3.4. The NLG setup consists of a template system that retrieves the response from a database based on the depression category value. The information shown in Fig. 4c is an example response. The release version of the app will feature targeted information based on collaboration with doctors and mental health care professionals.

5 Empirical Testing Results

To ensure the utmost care for user safety, initial testing was carried out against synthetic datasets, rather than human testers, who may be affected by issues arising from an app at early-stage development. The synthetic datasets were created from three logic trees shown in Fig. 5 designed to demonstrate three separate possible patterns of observable behaviour. Ten datasets were created from each tree, sized from 100 to 1000 records in 100-record multiples. Mindful was tasked with building a decision tree from each of the ten datasets for each of the three logic trees, creating 30 trees in total. In dataset generation, the probability of an attribute value taking either branch was set at 50%, with a 1% randomness factor added to limit tree accuracy to 99%.

Testing was performed on an Alcatel Pixi 3 (4.5) smartphone, purchased as a pre-paid device for $34.50. This phone features Android 5.1 operating system, an ARM Cortex A53-based MediaTek MT6735M quad-core processor and 1 GB of RAM.

5.1 Classification Accuracy

The results in Fig. 6 show that even with a dataset of just 100 records, the decision tree achieved a strict error rate of 11% or better across all three logic trees using exact PHQ score matching. However, if the accuracy is relaxed to also allow nearest neighbours (for example, PHQ-1 or PHQ-3 equates to PHQ-2), the error rate for Logic Tree 2

244   D. Yates and Md. Z. Islam

drops from 11% to just 3%. Given Mindful's role in alerting the user, the exact depression category may matter less than the fact that the user is alerted. Since records are created by the user submitting a PHQ-8 response, it would be ideal for classification accuracy to as high as possible, as soon as possible. Preliminary testing on 30-record datasets based on Logic Tree 1 delivered a strict accuracy of 76.6%, rising to 96.6% with relaxed 'nearest neighbour' accuracy. This suggests that even with only 30 days of records, broad states of mental health can be detected with good accuracy, but human trials are needed to confirm this. Moreover, other techniques such as Random Forest may achieve better results with fewer records, albeit at slower speeds.

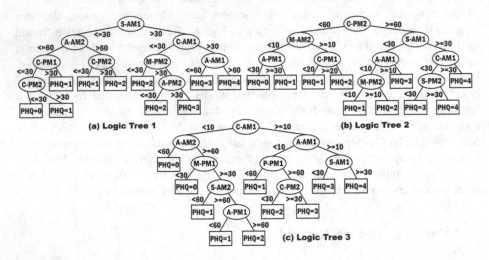

Fig. 5. The three logic trees from which synthetic datasets were drawn (node attributes A = accelerometer, C = call log, M = SMS log, P = proximity, S = screen, leaf nodes = PHQ category).

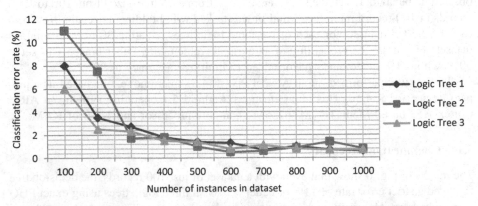

Fig. 6. Accuracy of J48 trees built by Mindful from datasets of the logic trees in Fig. 5.

5.2   Processing Speed

Build times for each decision tree were also recorded and in terms of speed, the Alcatel Pixi 3 phone was able to build a tree from a 1000-record/28-attribute dataset in under a second. The mean dataset record processing speed across all 30 datasets was 1342.5 records per second (SD = 92.42). Given our research has highlighted the possibility of a user's mental health state being detected with just a 30-record dataset, these speed results should provide sufficient encouragement that today's smartphones – even low-cost entry-level models – are fast enough to handle mobile data mining.

6   Discussion and Conclusion

Given the prevalence of smartphones and the close interaction many users have with them, the combination of smartphone and on-board machine learning creates a novel platform for exploring mental health. This paper has a number of opportunities for further research, including framework sample rates for the passive data acquisition and mental health state detection. The present rate of data acquisition is once per minute and the suggested rate of automated mental health state detection is one per hour, however, there is considerable scope for experimentation. Balancing optimum sample rates with maximising data granularity and smartphone battery life makes this an area of interest. The optimal PHQ-8 questionnaire submission frequency is another. Further afield, mobile data mining could be applied to anxiety detection and potentially other areas of mobile health. Alternate algorithms are also a noted area for research.

This paper has shown the readiness of smartphones for local data collection, pre-processing and data mining by presenting an example framework in the area of mental health. It has shown how sensor and log data can be combined with user input to create a training dataset for learning a model that can be used to predict mental health state changes. From a technology perspective, that this is achievable today reinforces the fact that smartphones are ready to perform data mining roles. But as such, this paper does not claim to be a finished product for improving mental health, only a first step. Future work will include human trials to better understand how Mindful can benefit patients. We invite colleagues in the mental health sciences to help progress this research to develop a smart mental health monitoring system.

References

1. Yates, D., Islam, Md.Z., Gao, J.: Implementation and performance analysis of data-mining classification algorithms on smartphones. In: Islam, R., et al. (eds.) AusDM 2018. CCIS, vol. 996, pp. 331–343. Springer, Singapore (2019). https://doi.org/10.1007/978-981-13-6661-1_26
2. Yates, D., Islam, M.Z., Gao, J.: DataLearner: a data mining and knowledge discovery tool for android smartphones and tablets. arXiv preprint arXiv:1906.03773 (2019)
3. Depression (n.d.). https://www.who.int/news-room/fact-sheets/detail/depression. Accessed 24 July 2019

4. Slade, T., Johnston, A., Oakley Browne, M.A., Andrews, G., Whiteford, H.: 2007 National Survey of Mental Health and Wellbeing: methods and key findings. Aust. N. Z. J. Psychiatry **43**(7), 594–605 (2009)
5. Jeong, T., Klabjan, D., Starren, J.: Predictive analytics using smartphone sensors for depressive episodes. arXiv preprint arXiv:1603.07692 (2016)
6. Shen, N., et al.: Finding a depression app: a review and content analysis of the depression app marketplace. JMIR mHealth uHealth **3**(1), e16 (2015)
7. Scherr, S., Goering, M.: Is a self-monitoring app for depression a good place for additional mental health information? Ecological momentary assessment of mental help information seeking among smartphone users. Health Commun. **77**, 1–9 (2019)
8. Arora, S., Yttri, J., Nilsen, W.: Privacy and security in mobile health (mHealth) research. Alcohol Res.: Curr. Rev. **36**(1), 143 (2014)
9. Bakker, D., Kazantzis, N., Rickwood, D., Rickard, N.: Mental health smartphone apps: review and evidence-based recommendations for future developments. JMIR Ment. Health **3** (1), e7 (2016)
10. Palmius, N., et al.: A multi-sensor monitoring system for objective mental health management in resource constrained environments (2014)
11. Lorant, V., Croux, C., Weich, S., Deliège, D., Mackenbach, J., Ansseau, M.: Depression and socio-economic risk factors: 7-year longitudinal population study. Br. J. Psychiatry **190**(4), 293–298 (2007)
12. Westoby, R., Westoby, R.: Mental Health, Housing and Homelessness: A Review of Issues and Current Practices, 2016 (2017)
13. Humphry, J.: Homeless and connected: mobile phones and the Internet in the lives of homeless Australians (2014)
14. Pash, C.: Smartphones have tightened their grip on Australian lives. https://www.businessinsider.com.au/smartphones-use-australia-deloitte-2018-11. Accessed 24 July 2019
15. Cornet, V.P., Holden, R.J.: Systematic review of smartphone-based passive sensing for health and wellbeing. J. Biomed. Inform. **77**, 120–132 (2018)
16. Chen, Z., et al.: Unobtrusive sleep monitoring using smartphones. In: Proceedings of the 7th International Conference on Pervasive Computing Technologies for Healthcare. ICST (Institute for Computer Sciences, Social-Informatics and Telecommunications Engineering) (2013)
17. Kroenke, K., Spitzer, R.L.: The PHQ-9: a new depression diagnostic and severity measure. Psychiatric Ann. **32**(9), 509–515 (2002)
18. Moriarty, A.S., Gilbody, S., McMillan, D., Manea, L.: Screening and case finding for major depressive disorder using the Patient Health Questionnaire (PHQ-9): a meta-analysis. Gen. Hosp. Psychiatry **37**(6), 567–576 (2015)
19. Fox, K.R.: The influence of physical activity on mental well-being. Public Health Nutr. **2** (3a), 411–418 (1999)
20. Ben-Zeev, D., Scherer, E.A., Wang, R., Xie, H., Campbell, A.T.: Next-generation psychiatric assessment: using smartphone sensors to monitor behavior and mental health. Psychiatr. Rehabil. J. **38**(3), 218 (2015)
21. Nezlek, J.B., Imbrie, M., Shean, G.D.: Depression and everyday social interaction. J. Pers. Soc. Psychol. **67**(6), 1101 (1994)
22. Quinlan, J.R.: C4. 5: Programs for Machine Learning. Elsevier (2014)
23. Breiman, L.: Random forests. Mach. Learn. **45**(1), 5–32 (2001)
24. Reiter, E., Dale, R.: Building applied natural language generation systems. Nat. Lang. Eng. **3** (1), 57–87 (1997)
25. Weka 3: Machine Learning Software in Java (n.d.). https://www.cs.waikato.ac.nz/ml/weka/. Accessed 24 July 2019

Industry Showcase

Predictive Analytics for Tertiary Learners in New Zealand Who Are at Risk of Dropping Out of Education

Wenying Xu[1], Scott Luo[1], Stephanie Hacksley[1], Tim Trewinnard[1],
Stuart Cambridge[2], and Syen Jien Nik[1(✉)] (iD)

[1] Jade Software Corporation, Christchurch, New Zealand
snik@jadeworld.com
[2] Tertiary Education Commission, Wellington, New Zealand

Abstract. This industry showcase covers a proof-of-concept predictive model in the education sector of New Zealand. Jade Software worked with New Zealand's Tertiary Education Commission on research to find out how to predict the likelihood of learners dropping out. Our model informs the implementation of intervention programs to support learners in completing their qualifications. The goal of this research is to identify a common data set across multiple types of tertiary education organizations and develop predictive models using the data set. We found that the Single Data Return is a viable data source to form a base model. By comparing the area under the receiver operator characteristic curve, we show that additional data sources, including the attendance data and the learner's results, are helpful in improving model performance. We also developed an interactive dashboard to facilitate estimating the return on investment for intervention programs and the optimal intervention threshold.

Keywords: Predictive modelling · Tertiary Education · Learner success

1 Background

The primary objective of the Jade Learner Retention Model (JLRM) is to reduce the levels of learner dropout in New Zealand's tertiary education organizations (TEOs). The model gives TEOs a solution that enables them to better direct their investment in intervention programs and reduce learner dropout rates. Implementing the JLRM enables TEOs to create targeted intervention programs for learners who are at greater risk of dropping out. This facilitates a higher success rate and a greater return on Investment (ROI) for intervention programs.

Jade Software and the Tertiary Education Commission (TEC) collaborated on research in order to:

- Analyze the TEO data to create a common data set. This enables the JLRM to be applicable across most TEOs in New Zealand.
- Evaluate the JLRM performance across multiple organizations.
- Assess the sensitivity of JLRM to additional data sets.

© Springer Nature Singapore Pte Ltd. 2019
T. D. Le et al. (Eds.): AusDM 2019, CCIS 1127, pp. 249–255, 2019.
https://doi.org/10.1007/978-981-15-1699-3_20

2 Research Approach

2.1 Defining Learner Dropout

We needed a meaningful definition of learner dropout that was consistent across multiple TEOs. For this research, we used the TEC's definition of a learner not completing their qualification within:

- 2 years for Level 1 to 3 qualifications
- 4 years for Level 4 to 7 non-degree qualifications
- 6 years for Level 7 and higher degree qualifications [1].

2.2 Defining Input Data and Predictors

To develop a common predictive model for most TEOs in New Zealand, we needed to find and use a consistent data set that was being captured by all TEOs.

Single Data Return (SDR)

The SDR was identified as the only common data set available. It's a set of data items that are specifically required by the Ministry of Education and the TEC for funding, monitoring performance, publishing performance information, and statistical reporting purposes [2]. TEOs in New Zealand are required to submit an SDR three times a year.

Using the SDR data, we developed up to 100 variables, including demographics and other measures. Some of these variables were extracted directly from the SDR data set. Three examples of the variables are equivalent full-time student value for the current qualification, highest secondary school qualification, and ethnicity code. Other variables were calculated and aggregated from data points within the SDR [3]. For example, percentage of course completion so far for the current qualification, number of credits passed so far for the current qualification, and total number of completed past qualifications.

To create training and evaluation data sets for the JLRM, we separated the timeline of the SDR data into four components. Figure 1 shows the construction of the data for Level 7 and higher qualifications. Learners whose latest qualification fall between 2009 and 2011 are included in the Training data set, while learners whose latest qualification fall between 2011 and 2012 are included in the Evaluation data set. The four defined components for Level 7 and higher qualifications are:

- Profile: 2007-2009. For every learner we calculated characteristics about their qualifications from the past two years. To achieve this, we left a two-year gap in the timeline between 2007 and 2009 so even the learners with a qualification at the beginning of 2009 would have the same two-year profile period.
- Training: 2009-2011. We used two years of data to train the JLRM. All learners with a qualification between 2009 and 2011 were used to train our model.
- Evaluation: 2011-2012. We evaluated our model by applying the trained model on the 2011 data. All learners with a qualification in 2011 were used to evaluate our model.

- Dropout: 2012-2018. We left a six-year gap in the timeline to allow for dropout to happen that the model needed to predict. Even learners with a qualification at the end of 2012 had the same six-year dropout period.

Profile	Training	Evaluation	Dropout
2007	2009	2011	2012 2018

Fig. 1. The data components for Level 7 and higher qualifications. We separated the timeline of the SDR data into four components to create training and evaluation data sets.

We only needed a four-year gap in the dropout component for Level 4 to 7 qualifications. This means the Profile, Training, and Evaluation components shifted two years closer to 2018 to include the latest available data. Similarly, the three components for Level 1 to 3 qualifications shifted by another two years closer to 2018.

Additional Data

To supplement the SDR data, we explored additional data that would be consistently available across multiple TEOs. Such data was limited due to the TEOs operating independently of each other outside the common need to submit the SDR data. Attendance and results data were the only two additional data sources that were considered to have potential consistency across multiple TEOs.

Data Combinations

Together with the TEC, we decided to use the data from four TEOs to train and evaluate models. The TEOs selected comprise one vocational college, one university, and two polytechnics. As per our non-disclosure agreement with the TEC, the names of the TEOs are not reported.

From these selected TEOs, three combinations of data were used to train and evaluate the JLRM:

- SDR data only for all four TEOs – This was to determine the model performance for more than one TEO using the standard SDR data set.
- SDR and attendance data for one TEO – This was to evaluate the benefit of using attendance data as an additional data source.
- SDR and results data for one TEO – This was to evaluate the benefit of using results data as an additional data source.

TEC provided SDR data for all four TEOs. We were also given attendance data from one TEO which we appended to their SDR data set. Similarly, we appended the results data from another TEO to their SDR data and evaluated the model performance. Table 1 shows the data sets we received from the four TEOs.

Table 1. Collection of data sources in addition to the SDR data we have for the four TEOs.

TEO (TEO type)	SDR	Attendance	Results
TEO1 (Vocational college)	Yes	Yes	No
TEO2 (Polytechnic)	Yes	No	Approximately 60% recorded
TEO3 (Polytechnic)	Yes	No	No
TEO4 (University)	Yes	No	No

Model Selection and Optimization

Using only SDR data, we built our models on the scikit-learn, h2o, TensorFlow, and R platforms. We explored the Random Forest, Gradient Boosting Machine, Extreme Gradient Boosting, Support Vector Machine, Deep Learning Classifier, and Logistic Regressions models.

We ran several iterations of data transformation and modelling. Gradient Boosting Machine on the h2o platform was confirmed as the preferred model approach. This is because it provided the optimal balance between computational speed, ease of use, ease of interpretation, and model performance.

Having selected the Gradient Boosting Machine, we tuned the hyperparameters in h2o using grid search to find optimal model performance. The hyperparameters we have in the grid search are number of trees, maximum tree depth, learning rate, learning rate reduction, row sampling rate, and column sampling rate [4].

We used cross-validation to quantify the model performance to make sure the model was not overfitted. We applied the cross-validated Gradient Boosting Machine on the Evaluation data set and reported the area under the Receiver Operating Characteristic curve (AUR).

We repeated the process of hyperparameters tuning, cross-validation, and evaluation for all the three data combinations listed above.

2.3 Model Results and Output

The AUR for all four TEOs using only the SDR data is 0.83. This AUR shows that the JLRM is a useful and valuable model for predicting learner dropout across several types of TEOs, using just the SDR data. The eight most important variables are percentage of course completion so far for the current qualification, percentage of credits failed so far for the current qualification, full time indicator, number of credits passed so far for the current qualification, which TEO a leaner belongs to, sum of tuition (FTE teaching) and vacation weeks for the current qualification, first character from Funding Category code in SDR, and self-learning hours required for the current qualification.

Including Additional Data

Using only the TEO1's data, adding learner attendance lifts the AUR of the model to 0.86. The AUR also improved slightly by adding the learner's results to the SDR data for TEO2. The improvement is smaller than the attendance data because only 60% of

TEO2's learners have results recorded against them. Table 2 shows the AUR for the three data combinations.

Table 2. This table shows the AUR for three data combinations. The AUR for all four TEOs using only the SDR data is 0.83. The AUR improved slightly by adding either learner attendance or learner's results to the SDR data.

Data set	AUR
Four TEOs with SDR only	0.83
TEO1 – SDR only, without attendance	0.80
TEO1 – With attendance	0.86
TEO2 – SDR only, without results	0.79
TEO2 – SDR only, with results	0.80

Return on Investment (ROI) of Intervention

Equation (1) is used to facilitate the TEC in estimating the ROI for the programs and the optimal intervention threshold:

$$ROI = \sum_k return * k * f_1 * f_2 - \left(running\ costs + \sum_l intervention\ cost * l \right) \quad (1)$$

where

- k is the total number of true positives included in the intervention threshold,
- *return* is the average value of retaining a learner,
- f_1 is the percentage likelihood of an intervention succeeding,
- f_2 is the percentage likelihood that a learner will take up an intervention,
- l is the total number of learners included in the intervention threshold,
- *running costs* include annual licence fees and marketing costs,
- *intervention cost* is the average cost of intervening with a learner.

These inputs enable the JLRM to indicate the intervention threshold that would provide the highest ROI. Alternatively, the intervention threshold can be guided by the allocated budget for interventions. Equation (1) can then be used to indicate the ROI at the budgeted intervention threshold.

For the ROI calculation to be accessible by the TEC, we developed a dashboard in Shiny that interacts directly with the Evaluation data set using the inputs above. We use interactive sliders to control the inputs above to correspond to the interventions, while k is given by the total number of true positives as predicted by the JLRM within the Evaluation data set. Figure 2 provides a screenshot of the graph outputs within the dashboard.

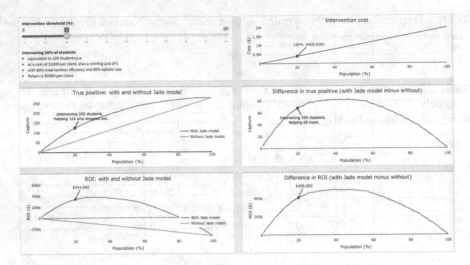

Fig. 2. A screenshot of the Shiny dashboard we created to facilitate the TEC estimating the ROI for the intervention, and the optimal intervention threshold.

3 Conclusions and Discussion

The JLRM provides a way for the TEC or individual TEOs to better target learners that are at a high risk of dropping out. An AUR of between 0.79 and 0.86 across multiple TEOs gives us confidence that the JLRM model would be effective in standardizing how intervention programs are targeted. Any of the task value, framing, or personal value intervention programs can be applied [5]. Alternatively, it can be a tiered approach - for example, less intensive interventions for mid- to high-risk learners and more intensive interventions for high-risk learners.

There are multiple options for implementing the JLRM into the operational process of a TEO or the TEC. When considering the options, we noted that:

- By using SDR data, the TEC can provide a multi-TEO service and direct the output of the JLRM to all, or a subset of, TEOs.
- Individual TEOs could adopt the model internally and improve model performance by including extra data sets.

In all scenarios, Jade would give TEC the output from JLRM, either through a self-service portal or an integration with an existing learner management system.

3.1 How Could the Model Be Improved?

Additional Data Sets

We showed that model performance improves with adding data that isn't present in the SDR data set. As part of the research, we looked at more data points collected by TEOs, including:

- Online course logins and results – These weren't used because they're too new and don't provide enough data to develop models.
- Campus Wi-Fi logs – These aren't consistently recorded or stored over time.

Consistently collecting data about online courses and Wi-Fi use can further improve the JLRM.

Cohort Targeting

The performance of the JLRM against multiple learner cohorts, compared to a single countrywide model, hasn't been assessed. Multiple models might outperform one model in the areas of TEO, level of qualification, and area of study.

However, using cohort-based models would create increased complexity. This is because the cohorts can overlap, thereby generating multiple risk scores for an individual learner.

Data Cadence

The JLRM used SDR data that was submitted at the end of each semester. Increasing the data cadence to run the model with data available during the semester might improve model performance. It can also reduce the time between a learner becoming at risk and an intervention being initiated.

References

1. Tertiary Education Commission: Educational Performance Indicators: Definitions and Methodology For Institutes of Technology and Polytechnics, Private Training Establishments, Universities and Wānanga Version 1.1
2. SDR. https://www.tec.govt.nz/funding/funding-and-performance/reporting/sdr/. Accessed 05 June 2019
3. Ministry of Education and Tertiary Education Commission: Single Data Return. A Manual for Tertiary Education Organisations and Student Management System Developers. Version 1.0 (2018)
4. h2o Gradient Boosting Machine. http://docs.h2o.ai/h2o/latest-stable/h2o-docs/data-science/gbm.html. Accessed 05 June 2019
5. Harackiewicz, J.M., Priniski, S.J.: Improving student outcomes in higher education: the science of targeted intervention. Annu. Rev. Psychol. **69**, 409–435 (2018)

Author Index

Printed in the United States
by Bookmasters

Printed in the United States
By Bookmasters